심선생 Math Series

심현성 지음
(Albert Shim)

5월 AP Calculus AB 시험대비
Final 총정리와 Practice Tests

심선생 Math Series

- Multiple Choice Questions 대비 18개 Topic과 61문제 총정리로 만점 받기
- Multiple Choice Questions 4 Full Tests
- Free Response Questions 6개 Topic 총정리로 만점 받기

저자 소개

심현성(Albert Shim) 선생님은

수능수학과 경시수학을 가르치다가 미국수학 전문가가 되었다. 레카스 아카데미를 거쳐 블루키프렙 대표이사 겸 대표강사를 지내다가 현재 TOP SEM학원의 대표이사 겸 Math 대표강사이기도 하다. 2008년 한국에서는 처음으로 "Math Level 2"를 출간 하였고 연이어 2009년에는 처음으로 "AP calculus"출간하였다. 현재는 10개국 이상의 나라에 Math관련 교재를 출간하고 있다. 특히, "Math Level 2 10 Practice Tests" , "AP Calculus AB&BC 핵심 편", "AP Calculus AB&BC 심화 편", "AMC10 & 12 특강"등은 미국 대학을 준비하는 거의 모든 학생들의 필수서적이 될 만큼 중요한 교재이며 베스트셀러 교재이기도 하다. 2008년부터 지금까지 10개국 이상에 출간한 책이 20권 이상이며 압구정에서 가장 많은 수강생을 가르치는 유명강사이다. 오프라인에서는 압구정에 위치한 TOP SEM학원에서 강의하고 있으며 온라인에서는 SAT,AP,IB 즉 미국대학 입시 전문 인터넷 동영상 강의 전문업체인 마스터프렙(www.masterprep.net)에서 Math의 거의 모든 분야를 강의하고 있으며 해당 사이트에서도 No.1 수학강사로 차별성 있는 톡톡 뛰는 강의로 정평이 나 있다.

수업문의

TOP SEM학원 (02-511-4235, www.topsem.co.kr)
인터넷 동영상업체 마스터프렙 (www.masterprep.net)

Multiple Choice Questions (MCQ)

두 개의 파트로 구성이 되어 있고

Part A는 30문제 60분이 주어지며 계산기를 사용할 수 없다.

Part B는 15문제 45분이 주어지며 계산기 사용이 가능하다.

Free Response Questions (FRQ)

두 개의 파트로 구성이 되어 있고

Part A는 2문제 30분이 주어지며 계산기 사용이 가능하다.

Part B는 4문제 60분이 주어지며 계산기 사용이 불가능하다.

대부분의 학생들은 어떻게 느끼는가?

MCQ의 경우에는 Part A가 쉽고 시간이 넉넉하며 FRQ의 경우에는 Part B가 쉽게 느껴진다고 한다. MCQ, FRQ 모두 오히려 계산기 사용이 가능한 파트를 더 어렵게 느낀다는 것이 대부분 학생들의 답변이다.

1. Multiple Choice Questions를 위한 18개의 핵심 Topic과 61개의 핵심 문제를 제공함으로써 단기간에 AP Calculus AB를 준비할 수 있게 하였다.

2. AP Calculus AB Full Test 4회분과 자세한 해설을 실었으며 Quality 높은 문제들로 구성되어 있다.

3. Free Response Questions를 위한 6개의 Topic과 심선생만의 차별화된 접근 방식을 제공하였다.

4. 그 어느 교재에서도 볼 수 없는 심선생만의 차별화된 강의와 해설이 그대로 담겨 있다.

5. 최단기간 준비로 만점을 받을 수 있게 구성하였다.

이 책의
활용 방법

1. Free Response Questions의 경우 본 교재를 모두 풀어봤다 하더라도 Collegeboard에서 제공하는 최근 기출문제 3~5년 분량은 반드시 풀어보자.

2. FRQ 서술형의 경우 Collegeboard에서 제공하는 해설과 다른 방식으로 써도 무관하다. 즉, Collegeboard에서 제공하는 해설지가 만능은 아니다!

3. 4개의 Full Test를 풀어보기 전에 반드시 MCQ 18개의 Topic과 61개의 핵심문제를 완성하도록 한다.

1. 학교에서 AP Calculus를 수강했던 학생들만 5월 AP Calculus 시험에 응시하자!

학교에서 배우기도 전에 시험을 봐 버린다면 미국 내의 고교 교사들은 정작 중요한
AP Calculus 수업 수강을 못하게 하는 경우가 많다. 이는 대학 입시에도 좋지 못한
경우가 된다.

2. 시간을 절약하자!

한국의 입시에서는 수학이 가장 중요한 과목 중 하나이지만 미국의 대학 입시에서는
아주 일부 중 하나이다. 이 책 저 책 많이 사서 쩔쩔매는 학생들을 종종 만나는데,
전혀 그럴 필요가 없다. 왜냐고? AP Calculus 시험은 어느 정도만 풀어도 만점인
시험이기 때문이다.
다른 Science 과목이나 Social 과목들보다 만점을 많이 주는 시험이다. 그러니,
다른 과목에 좀 더 많은 시간을 투자하여야 한다.

3. 강박증을 버리자!

한국의 수능 수학에서는 1개만 틀려도 대학의 합격 여부가 결정되지만, 앞에서도
말한 것처럼 AP Calculus 시험은 어느 정도 기준만 통과하면 만점인 시험이다.
그러니, 시험장에서도 자신 있는 것 먼저 풀고 MCQ의 경우 모르는 문제는 찍으면 된다.
틀려도 감점은 없다. FRQ의 경우에도 자신 있는 문제들만 잘 골라 써도 만점이니
너무 강박관념을 가지고 시험을 보지 않아도 된다.
일부 학생들이 시험을 보고 난 후 결과도 나오지 않았는데 성적이 안 나올 것 같다며
취소하려 해서 여러 번 말렸었다. 그런데 이게 웬일인가? 그 학생들 모두 만점이
나왔었다. 그러니, 좀 못 푼 문제가 있다고 해서 실망할 필요는 없는 것이다.

Preface

필자가 AP Calculus 책을 집필한 지도 어느덧 10년의 세월이 흘렀다. 강의 현장에서 느낀
점들을 하나도 빠짐없이 독자들에게 전달해 주고 싶은 마음으로 집필을 해 왔다.
이 책은 5월 AP Calculus AB를 준비하는 학생들에게 단기간에 최대의 효과를 보게 하려고
집필하였다. 방대하게 보이는 AP Calculus AB 시험을 간결하게 정리하고 싶었고
무엇보다도 간결함 속에서도 필자만의 노하우만큼은 빠지지 않게 정리하여 전달하고
싶었다. AP Calculus 실전 문제를 2007년에 처음으로 개발하여 학생들에게 풀려봤었고
매년 모의고사를 개발하고 수업을 하면서 그리고 학생들의 반응과 결과를 보면서 매년
수정에 수정을 거듭했다.

그동안 많은 학생들이 이 책의 출판에 대한 요구의 목소리도 높았다. 누구나 쉽게 접근하고
간결하게 시험을 준비할 수 있는… 그 어느 책보다 질 좋은 문제들만 담고 그 어느 책에서도
볼 수 없는 해설을 실은 책을 만들고 싶었고 이제 그 꿈을 이룬 것 같아 상당히 기쁘다.
이 책의 집필 과정은 상당히 고되고 힘들었다. 그동안 한 인간으로서 할 수 있는 모든 노력은
다 했다. 앞으로도 본인이 할 수 있는 모든 노력을 다할 것을 약속드린다.

그동안 많은 분들의 도움이 있었다.
필자의 학원 운영을 책임지시는 현수현 원장님, 필자의 온라인 강의를 허락해 주신
마스터프렙 권주근 대표님, 무엇보다도 소중한 자녀를 필자에게 맡겨주신 학부모님들께
지면을 빌려 감사한 마음을 전한다.
이 책이 학생들에게는 없어서는 안 될 길잡이가 되기를 간절히 바라는 바이다.

2019. 03. 05
심현성

Contents

Multiple Choice Questions
-핵심내용 18 Topics와 61문제로 총정리 하기

TOPIC 1 – Limit 계산   P.10

TOPIC 2 – L'Hopital's Rule   P.13

TOPIC 3 – Asymptote   P.16

TOPIC 4 - $\lim\limits_{x\to\infty} f(x)$의 해석   P.19

TOPIC 5 – The Definition of Derivative   P.26

TOPIC 6 – Derivative 계산   P.31

TOPIC 7 – The derivative of inverse function   P.38

TOPIC 8 – Integral 계산   P.41

TOPIC 9 – Graph   P.52

TOPIC 10 – Related Rates   P.63

TOPIC 11 – Motion   P.70

TOPIC 12 – Area & Volume   P.74

TOPIC 13 – Differential Equation   P.80

TOPIC 14 – T.D.M.L   P.83

TOPIC 15 – Line Tangent Equation   P.87

TOPIC 16 - $\lim\limits_{n\to\infty}\sum\limits_{k=1}^{n} f(k) = \int_{a}^{b} f(x)dx$ P.90

TOPIC 17 – Riemann Sum   P.99

TOPIC 18 – Slope Field   P.104

Practice Test 1   P.107

Practice Test 2   P.157

Practice Test 3   P.207

Practice Test 4   P.255

Free Response Questions
-6 Topics로 만점 받기

TOPIC 1 – Graph   P.301

TOPIC 2 – Related Rates   P.311

TOPIC 3 – Table   P.321

TOPIC 4 – Area & Volume   P.331

TOPIC 5 – Motion   P.349

TOPIC 6 – Differential Equation   P.355

Multiple Choice Questions

 핵심내용 18 Topics와 61문제로 총정리하기

 시작하기 전에 …

AP Calculus AB 시험을 준비하는 데 있어서 Multiple Choice Question에 꼭 필요한 내용과 문제들만 모아 두었다. 실전 문제에 들어가기에 앞서서 반드시 알아야 할 내용과 문제들만 실었으니 꼼꼼하게 공부하도록 해야 한다.

Topic 1 - Limit 계산

1. $\dfrac{\infty}{\infty}$

$$\lim_{x \to \infty} \frac{cx^m + dx^{m-1}}{ax^n + bx^{n-1} + \cdots} = \frac{c}{a} \qquad (m = n)$$

$$= \infty \qquad (m > n)$$

$$= 0 \qquad (m < n)$$

2. $\displaystyle\lim_{x \to a} f(x)$ $\quad \begin{cases} \displaystyle\lim_{x \to a^-} f(x) \\ \displaystyle\lim_{x \to a^+} f(x) \end{cases}$

즉, Left Hand Limit과 Right Hand Limit으로 나누어 계산한 후 두 계산 결과가 같을 때 $\displaystyle\lim_{x \to a} f(x)$ 값이 Exist.

1. $\lim\limits_{x \to \infty} \dfrac{2x-1}{\sqrt{x^2+3}} =$

   (A) $-\infty$ (B) 0 (C) 2 (D) ∞

2. $\lim\limits_{x \to 3} \dfrac{x-3}{x^2-2x-3} =$

   (A) $\dfrac{1}{4}$ (B) $\dfrac{1}{2}$ (C) 1 (D) 2

3. $\lim\limits_{x \to 0} \dfrac{|x|}{x} =$

   (A) 0 (B) 1 (C) 2 (D) Nonexistent

4. $\lim\limits_{x \to 4} \dfrac{\sqrt{x}-2}{x-4} =$

   (A) $\dfrac{1}{4}$ (B) $\dfrac{1}{2}$ (C) 0 (D) 2

Solution

1. (C)

$$\lim_{x \to \infty} \frac{2 - \dfrac{1}{x}}{\sqrt{1 + \dfrac{3}{x^2}}} = 2$$

2. (A)

$$\lim_{x \to 3} \frac{x - 3}{(x - 3)(x + 1)} \ \Rightarrow \ \lim_{x \to 3} \frac{1}{x + 1}$$

$$\Rightarrow \ \lim_{x \to 3-} \frac{1}{x + 1} = \frac{1}{4}, \ \lim_{x \to 3+} \frac{1}{x + 1} = \frac{1}{4}$$

3. (D)

$$\lim_{x \to 0+} \frac{x}{x} = 1, \ \lim_{x \to 0-} \frac{-x}{x} = -1$$

4. (A)

$$\lim_{x \to 4} \frac{\sqrt{x} - 2}{\sqrt{x}^2 - 2^2} \ \Rightarrow \ \lim_{x \to 4} \frac{\sqrt{x} - 2}{(\sqrt{x} - 2)(\sqrt{x} + 2)} \ \Rightarrow \ \lim_{x \to 4} \frac{1}{\sqrt{x} + 2}$$

$$\lim_{x \to 4+} \frac{1}{\sqrt{x} + 2} = \frac{1}{4}, \ \lim_{x \to 4-} \frac{1}{\sqrt{x} + 2} = \frac{1}{4}$$

Topic 2 - L'Hopital's Rule

Limit을 계산할 때, $\dfrac{\infty}{\infty}$ 또는 $\dfrac{0}{0}$ 형태에 대해서는 다음과 같이 Denominator, Numerator를 각각 Differentiate 한다.

$\Rightarrow$ Denominator 또는 Numerator 둘 중 하나가 0 또는 ∞가 아닐 때까지 Differentiate 한다.

Exercise

5. $\lim\limits_{x \to 0} \dfrac{e^x + \cos x - 2}{x^2 - x} =$

(A) -2 (B) -1 (C) 0 (D) $\dfrac{1}{2}$

6. $\lim\limits_{x \to 0} \dfrac{2\tan x}{e^{5x} - 1} =$

(A) 0 (B) $\dfrac{2}{5}$ (C) $\dfrac{4}{5}$ (D) 2

7. $\lim\limits_{x \to \infty} \dfrac{x^2}{e^x} =$

(A) 0 (B) 1 (C) e (D) 2

x	$f(x)$	$f'(x)$	$f''(x)$	$f^{(3)}(x)$
5	0	0	0	2

8. The third derivative of the function f is continuous on the interval $(0, 4)$. Values for f and its first three derivatives at $x = 5$ are given in the table above. What is $\lim\limits_{x \to 5} \dfrac{f(x)}{(x-5)^3}$?

(A) 0 (B) $\dfrac{1}{6}$ (C) $\dfrac{1}{3}$ (D) Nonexistent

Solution

5. (B)

L'Hopital's Rule!

$\lim\limits_{x \to 0} \dfrac{e^x + \cos x - 2}{x^2 - x} \Rightarrow \dfrac{0}{0}$ 모양!

$\Rightarrow \lim\limits_{x \to 0} \dfrac{(e^x + \cos x - 2)'}{(x^2 - x)'} \Rightarrow \lim\limits_{x \to 0} \dfrac{e^x - \sin x}{2x - 1} = \dfrac{1}{-1} = -1$

6. (B)

L'Hopital's Rule!

$\lim\limits_{x \to 0} \dfrac{2\tan x}{e^{5x} - 1} \Rightarrow \dfrac{0}{0}$ 모양!

$\Rightarrow \lim\limits_{x \to 0} \dfrac{(2\tan x)'}{(e^{5x} - 1)'} \Rightarrow \lim\limits_{x \to 0} \dfrac{2\sec^2 x}{5e^{5x}} = \dfrac{2}{5}$

7. (A)

L'Hopital's Rule!

$\lim\limits_{x \to \infty} \dfrac{x^2}{e^x} \Rightarrow \dfrac{\infty}{\infty}$ 모양!

$\Rightarrow \lim\limits_{x \to \infty} \dfrac{(x^2)'}{(e^x)'} = \lim\limits_{x \to \infty} \dfrac{2x}{e^x} \Rightarrow$ L'Hopital's Rule! $\Rightarrow \lim\limits_{x \to \infty} \dfrac{(2x)'}{(e^x)'} = \lim\limits_{x \to \infty} \dfrac{2}{e^x} = \dfrac{2}{\infty} = 0$

8. (C)

$\lim\limits_{x \to 5} \dfrac{f(x)}{(x-5)^3} \Rightarrow \dfrac{0}{0}$ 모양!

$\Rightarrow \lim\limits_{x \to 5} \dfrac{f'(x)}{3(x-5)^2} \Rightarrow \dfrac{0}{0}$ 모양!

$\Rightarrow \lim\limits_{x \to 5} \dfrac{f''(x)}{6(x-5)} \Rightarrow \dfrac{0}{0}$ 모양!

$\Rightarrow \lim\limits_{x \to 5} \dfrac{f'''(x)}{6} = \dfrac{2}{6} = \dfrac{1}{3}$

Topic 3 - Asymptote

$y = \dfrac{g(x)}{f(x)}$ 에 대해…

1. Vertical Asymptote : $f(x) = 0$인 x값

2. Horizontal Asymptote : $\displaystyle\lim_{x \to \infty} \dfrac{g(x)}{f(x)}$와 $\displaystyle\lim_{x \to -\infty} \dfrac{g(x)}{f(x)}$ 계산

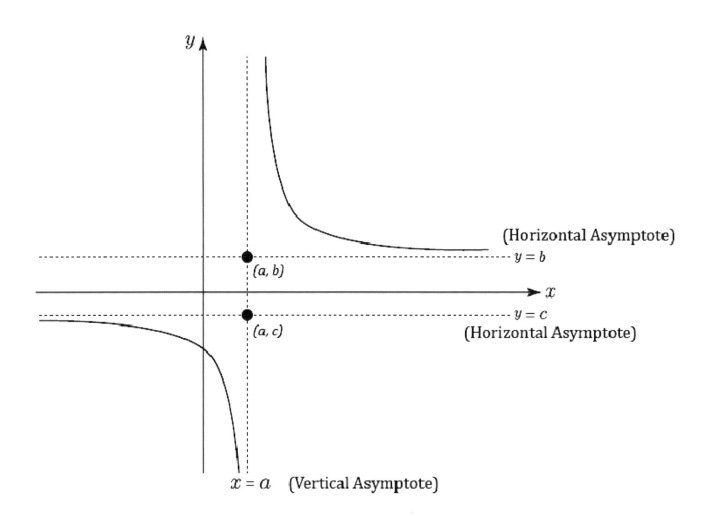

⟨?⟩ Exercise

9. What is(are) horizontal asymptote(s) of the graph of $y = \dfrac{10 + 3^x}{3 - 3^x}$ on the xy-plane?

(A) $y = -1$

(B) $y = 0$

(C) $y = -1$ and $\dfrac{10}{3}$

(D) $y = \dfrac{10}{3}$

10. Which of the following lines is(are) asymptote(s) of the graph of $f(x) = \dfrac{x^2 - 3x - 2}{x^2 - 4}$?

| I. $y = 1$ | II. $x = \pm 2$ | III. $x = -1$ | IV. $x = -2$ and -1 |

(A) I only (B) II only (C) I and II only (D) II and IV only

Solution

9. (C)

- $\displaystyle \lim_{x \to \infty} \frac{\dfrac{10}{3^x}+1}{\dfrac{3}{3^x}-1} = \frac{0+1}{0-1} = -1$

- $\displaystyle \lim_{x \to -\infty} \frac{10+3^x}{3-3^x} = \frac{10+3^{-\infty}}{3-3^{-\infty}} = \frac{10+\dfrac{1}{3^{\infty}}}{3-\dfrac{1}{3^{\infty}}} = \frac{10}{3}$

10. (C)

① Vertical Asymptote : $x^2 - 4 = 0 \Rightarrow x = \pm 2$

② Horizontal Asymptote :
- $\displaystyle \lim_{x \to \infty} \frac{x^2 - 3x - 2}{x^2 - 4} = 1$
- $\displaystyle \lim_{x \to -\infty} \frac{x^2 - 3x - 2}{x^2 - 4} = 1$

Therefore, $x = \pm 2$, $y = 1$

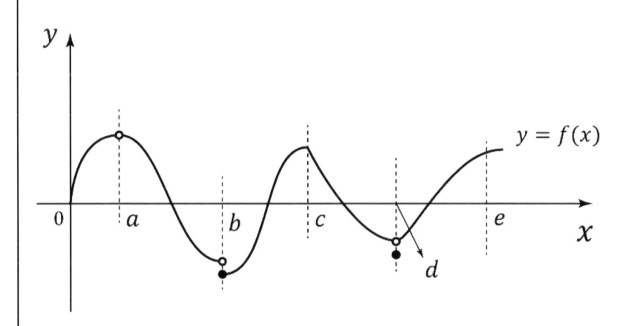

	a	b	c	d	e
• Continuity	X	X	O	X	O
• Limit exist	O	X	O	O	O
• Function exist	X	O	O	O	O
• Differentiable	X	X	X	X	O

위의 그림과 Table로부터 다음의 ① ~ ⑧까지를 알 수 있다.

① Continuity
　　⇒ Limit exist ⇒ Function exist
　　⇒ Differentiable은 알 수 없음

② Limit exist
　　⇒ Continuity, Function exist, Differentiable 모두 확신할 수 없음

③ Differentiable
　　⇒ Continuity ⇒ Limit exist ⇒ Function exist

④ Removable discontinuity $(a, 0)$, $(d, 0)$

$\lim\limits_{x \to a-} f(x) = \lim\limits_{x \to a+} f(x)$이지만 Discontinuity

⑤ Jump discontinuity $(b, 0)$

$\lim\limits_{x \to a-} f(x) \neq \lim\limits_{x \to a+} f(x)$이고 Discontinuity

⑥ The Definition of Continuity

$$\boxed{\lim_{x \to a-} f(x) = \lim_{x \to a+} f(x) = f(a)}$$

$$= \boxed{\lim_{x \to a} f(x) = f(a)}$$

⑦ $\lim\limits_{x \to a} f(x)$ exist

$\Rightarrow \lim\limits_{x \to a-} f(x) = \lim\limits_{x \to a+} f(x)$ or

$\Rightarrow \lim\limits_{x \to a} f(x) = c$ (※$c = f(a)$ or $c \neq f(x)$)

⑧ Differentiability

$y = f(x)$

②

①

$x \quad x$

a

$\Rightarrow$ ① $\lim\limits_{x \to a-} \dfrac{f(x) - f(a)}{x - a}$ = ② $\lim\limits_{x \to a+} \dfrac{f(x) - f(a)}{x - a}$

(Example) 미분불능 (Not Differentiable)의 예

①

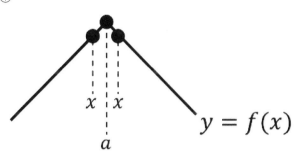

$$\Rightarrow \lim_{x \to a-} \frac{f(x)-f(a)}{x-a} \neq \lim_{x \to a+} \frac{f(x)-f(a)}{x-a}$$

②

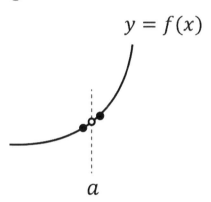

$y = f(x)$

$$\Rightarrow \lim_{x \to a-} \frac{f(x)-f(a)}{x-a} \ : \ f(a) \text{ does not exist}$$

$$\Rightarrow \lim_{x \to a+} \frac{f(x)-f(a)}{x-a} \ : \ f(a) \text{ does not exist}$$

③

$y = f(x)$

$$\Rightarrow \lim_{x \to a-} f(x) \neq \lim_{x \to a+} f(x) \text{ 이므로 } \lim_{x \to a-} \frac{f(x)-f(a)}{x-a} \neq \lim_{x \to a+} \frac{f(x)-f(a)}{x-a}$$

11. If $\begin{cases} f(x) = \dfrac{x^2 - 3x + 2}{x - 1} & (x \neq 1) \\ f(1) = k \end{cases}$, and if f is continuous at $x = 1$, then $k =$

(A) -1 (B) 0   (C) 1   (D) 2

12. If $\lim\limits_{x \to 5} f(x) = 3$, which of the following must be true?

I. $f(5) = 3$
II. $f'(5) = 3$
III. f is continuous at $x = 5$

(A) None   (B) I only   (C) II only   (D) I and III only

13. At $x = 2$, the function given by $f(x) = \begin{cases} x^2 + 1, & x \geq 2 \\ 4x - 3 & x < 2 \end{cases}$ is

(A) neither continuous nor differentiable

(B) continuous but not differentiable

(C) differentiable but not continuous

(D) both continuous and differentiable

14. The function $f(x) = \begin{cases} 2x & (x \leq 1) \\ e^x & (x > 1) \end{cases}$

(A) is continuous everywhere.

(B) is differentiable at $x = 1$.

(C) is continuous but not differentiable at $x = 1$.

(D) has a jump discontinuity at $x = 1$.

15. The function $f(x) = \begin{cases} \dfrac{x^2 - 2x - 3}{x - 3} & (x \neq 3) \\ 1 & (x = 3) \end{cases}$

(A) is continuous everywhere.

(B) is differentiable at $x = 3$.

(C) is continuous but not differentiable at $x = 3$.

(D) has a removable discontinuity at $x = 3$.

Solution

11. (A)

$$\lim_{x \to 1} \frac{x^2 - 3x + 2}{x - 1} = f(1) = k$$

$$\Rightarrow \lim_{x \to 1} \frac{(x-1)(x-2)}{x-1} = f(1) = k$$

$$\Rightarrow \lim_{x \to 1}(x-2) = k \Rightarrow \lim_{x \to 1-}(x-2) = -1 = k, \ \lim_{x \to 1+}(x-2) = -1 = k$$

12. (A)

$\lim\limits_{x \to 5} f(x) = 3$은 $\lim\limits_{x \to 5-} f(x) = \lim\limits_{x \to 5+} f(x)$를 의미한다.

즉, $\lim\limits_{x \to 5} f(x)$ exists.

$\lim\limits_{x \to 5} f(x)$ exist 한다고 하여

$x = 5$에서 continuous, $f(5)$ exist 여부, $x = 5$에서 differentiable 여부는 확실히 알 수 없다.

13. (D)

- $\lim\limits_{x \to 2-} \dfrac{f(x) - f(2)}{x - 2} \Rightarrow \lim\limits_{x \to 2-} \dfrac{4x - 3 - 5}{x - 2} = \lim\limits_{x \to 2-} \dfrac{4(x - 2)}{(x - 2)} = 4$

- $\lim\limits_{x \to 2+} \dfrac{f(x) - f(2)}{x - 2} \Rightarrow \lim\limits_{x \to 2+} \dfrac{x^2 + 1 - 5}{x - 2} = \lim\limits_{x \to 2+} \dfrac{(x - 2)(x + 2)}{x - 2} = 4$

$\lim\limits_{x \to 2-} \dfrac{f(x) - f(2)}{x - 2} = \lim\limits_{x \to 2+} \dfrac{f(x) - f(2)}{x - 2}$ 이므로 function f는 $x = 2$에서 differentiable!

or

$f'(x) = \begin{cases} (x^2 + 1)' = 4 \\ (4x - 3)' = 4 \end{cases}$ 이므로 differentiable

$\Rightarrow$ Differentiable이면

$\Rightarrow$ function f가 $x = 2$에서 differentiable이면 반드시 continuous이고

$\lim\limits_{x \to 2} f(x)$도 exist 하여 $f(2)$도 exist 한다.

14. (D)

function f는 $x=1$에서 discontinuity이고

$\lim\limits_{x\to 1+} e^x \neq \lim\limits_{x\to 1-} 2x$ 이므로 $x=1$에서 jump discontinuity.

15. (D)

function f는 $x=3$에서 discontinuity이고

$\lim\limits_{x\to 3-} f(x) = \lim\limits_{x\to 3+} f(x)$ 이므로 removable discontinuity.

Topic 5 - The Definition of Derivative

①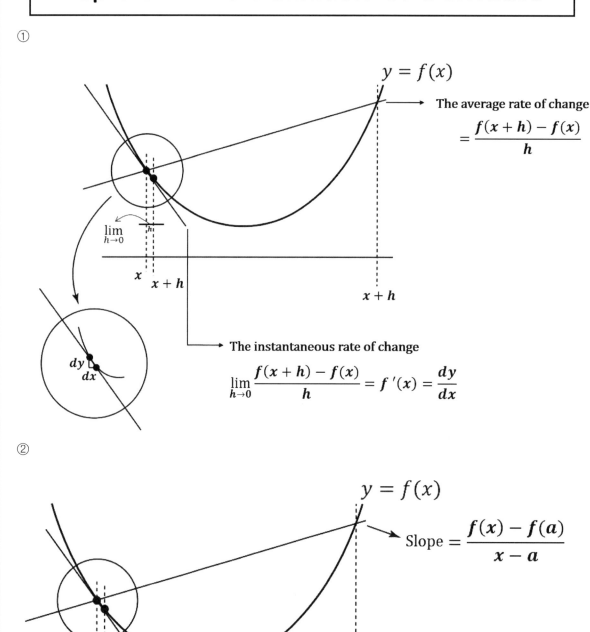

The average rate of change

$$= \frac{f(x+h) - f(x)}{h}$$

$y = f(x)$

$$\lim_{h \to 0}$$

The instantaneous rate of change

$$\lim_{h \to 0} \frac{f(x+h) - f(x)}{h} = f'(x) = \frac{dy}{dx}$$

②

$$y = f(x)$$

$$\text{Slope} = \frac{f(x) - f(a)}{x - a}$$

$$\text{Slope} = \lim_{x \to a} \frac{f(x) - f(a)}{x - a} = f'(a)$$

Exercise

16. Let f be a function such that $\lim\limits_{h \to 0} \dfrac{f(e+h)-f(e)}{h} = 3$, which of the following must be true?

> I. The derivative of f is continuous at $x=e$.
> II. $f'(e)=3$
> III. $\lim\limits_{x \to e} f(x) = 3$

(A) I only   (B) II only   (C) I and II only   (D) II and III only

17. If f is a function such that $\lim\limits_{x \to 3} \dfrac{f(x)-f(3)}{x-3} = 3$, which of the following is not true?

(A) $\lim\limits_{x \to 3} f(x)$ exists.

(B) $f(3)$ exists.

(C) f is continuous at $x=3$.

(D) $f(3)=3$

18. $\lim\limits_{x \to 0} \dfrac{\cos x - 1}{x}$ is

(A) $f'(0)$, where $f(x)=\sin x$

(B) $f'(1)$, where $f(x)=\cos x$

(C) $f'(0)$, where $f(x)=\cos x$

(D) $f'(1)$, where $f(x)=\cos(x+1)$

Solution

16. (B)

$\lim\limits_{h \to 0} \dfrac{f(x+h)-f(x)}{h} = f'(x)$에서 $x = e$이므로 $f'(e) = 3$.

즉, $y = f(x)$는 $x = e$에서 differentiable이므로

I. f'이 $x = e$에서 continuous 여부는 알 수 없다.

II. $f'(e) = 3$이고,

III. $\lim\limits_{x \to e} f(x)$이 exist 한다.

정답은, (B).

* $f'(e) = 3$은 $x = e$에서 The Slope of Line tangent가 3이라는 의미이지 $\lim\limits_{x \to e} f(x) = 3$이라고 확신할 수는 없다.

(Example)

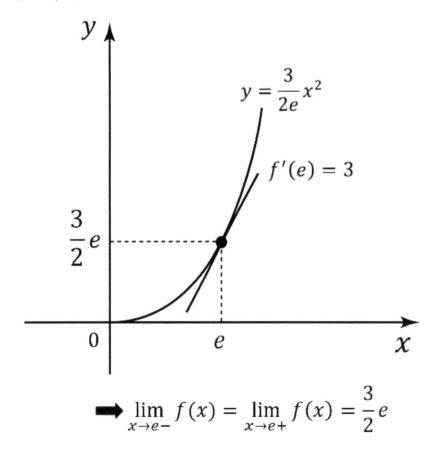

$$\Rightarrow \lim_{x \to e-} f(x) = \lim_{x \to e+} f(x) = \frac{3}{2}e$$

다음의 경우를 보자.

(Example 1)

(Example 2)

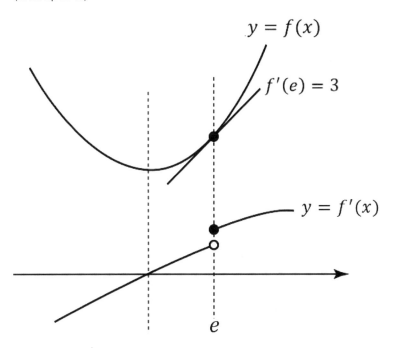

위의 (Example 1), (Example 2)의 경우에서 보는 것처럼 $f'(e)$가 exist 하더라도 f' graph가 $x = e$에서 continuous 할 수도 있고 아닐 수도 있다.

17. (D)

$$\lim_{x \to a} \frac{f(x) - f(a)}{x - a} = f'(a) \text{ 에서 } a = 3 \text{이므로 } f'(3) = 3.$$

즉, function f는 $x = 3$에서 continuous이고 $f(3)$도 exist하고 $\lim_{x \to 3} f(x)$도 exist 한다.

(D)번의 $f(3) = 3$인지는 확신할 수 없다. 예를 들어, 다음과 같은 경우가 생길 수 있기 때문이다.

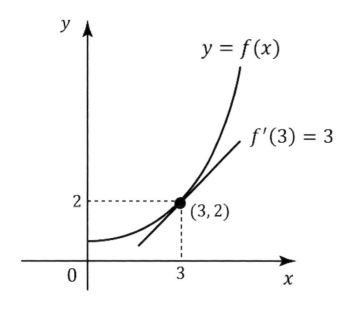

즉, $x = 3$에서 the slope of line tangent가 3인 것이지 $f(3) = 3$인 것이 아니기 때문이다.

18. (C)

$$\lim_{x \to a} \frac{f(x) - f(a)}{x - a} = f'(a) \text{에서 } f(x) = \cos x \text{이고 } a = 0 \text{이므로}$$

$f(x) = \cos x$에서 $f'(0)$을 구하는 것이다.

Topic 6 - Derivative 계산

<u>Derivative 계산법</u>

1. 기본적인 Formula

Formula (1)	$\bullet\ y = x^n\ \Rightarrow\ y' = n \cdot x^{n-1}$
	$\bullet\ y = f(x) \cdot g(x)\ \Rightarrow\ y' = f'(x) \cdot g(x) + f(x) \cdot g'(x)$
	$\bullet\ y = f(x) \pm g(x)\ \Rightarrow\ y' = f'(x) \pm g'(x)$
	$\bullet\ y = c \cdot f(x)\ \Rightarrow\ y' = c \cdot f'(x)\ (c \neq 0)$
	$\bullet\ y = \dfrac{g(x)}{f(x)}\ \Rightarrow\ y' = \dfrac{g'(x) \cdot f(x) - g(x) \cdot f'(x)}{(f(x))^2}$ (단, $f(x) \neq 0$)
	$\bullet\ y = c\ \Rightarrow\ y' = 0$

Formula (2)

① 삼각함수의 미분

$\Rightarrow$ $\boxed{c\text{로 시작} \rightarrow\ -\csc(\text{단, } \sin x,\ \cos x \text{ 제외}),\ t\text{ 포함} \rightarrow (\ )^2}$

- $(\sin x)' \rightarrow \cos x$ (※ $\sin x$와 $\cos x$는 서로 주고받는다.)
- $(\cos x)' \rightarrow -\sin x$
- $(\tan x)' \rightarrow \sec^2 x$ (타면 →시켜멓다) (t포함 → $(\ )^2$)
- $(\sec x)' \rightarrow \sec x \tan x$ (시커멓다 → 석탄!)
- $(\cot x)' \rightarrow -\csc^2 x$
- $(\csc x)' \rightarrow -\csc x \cot x$ (코 시커먼 것은 → 코 탄 것이다!)

② 역삼각함수의 미분

$\Rightarrow$ $\boxed{\begin{array}{l} c\text{로 시작하면 Negative!} \\ \sin^{-1}x,\ \tan^{-1}x,\ \sec^{-1}x\ \text{만 암기하면} \\ \cos^{-1}x,\ \cot^{-1}x,\ \csc^{-1}x\text{는 앞에 }(-)\text{만 붙는다.} \end{array}}$

- $(\sin^{-1}x)' \rightarrow \dfrac{1}{\sqrt{1-x^2}}(-1 < x < 1)$ 　　　$\bullet\ (\tan^{-1}x)' \rightarrow \dfrac{1}{1+x^2}$

- $(\cos^{-1}x)' \rightarrow -\dfrac{1}{\sqrt{1-x^2}}(-1 < x < 1)$ 　　$\bullet\ (\cot^{-1}x)' \rightarrow -\dfrac{1}{1+x^2}$

- $(\sec^{-1}x)' \rightarrow \dfrac{1}{|x|\sqrt{x^2-1}}(|x| > 1)$

- $(\csc^{-1}x)' \rightarrow -\dfrac{1}{|x|\sqrt{x^2-1}}(|x| > 1)$

③ 그 외의 공식들

- $(\log_a x)' \rightarrow \dfrac{1}{x \ln a}$ 　　　$\bullet\ (a^x)' \rightarrow a^x \ln a$ 　　　$\bullet\ (\ln x)' \rightarrow \dfrac{1}{x}$ 　　　$\bullet\ (e^x)' \rightarrow e^x$

2. Chain Rule

6가지 Chain Rule

ⓐ $y = (\bullet)^n \Rightarrow y' = n(\bullet)^{n-1} \cdot \bullet'$

(Ex) $y = (3x^2 + 5x - 1)^3 \Rightarrow y' = 3(3x^2 + 5x - 1)^{3-1} \cdot (6x + 5)$

ⓑ $y = \sin\bullet \Rightarrow y' = \bullet' \cdot \cos\bullet$

(Ex) $y = \sec(5x + 1) \Rightarrow y' = \sec(5x + 1)cdottan(5x + 1) \times 5$

ⓒ $y = \sin^{-1}\bullet \Rightarrow y' = \dfrac{1}{\sqrt{1 - \bullet^2}} \times \bullet'$

(Ex) $y = \tan^{-1}(5x^2) \Rightarrow y' = \dfrac{1}{1 + (5x^2)^2} \times 10x$

ⓓ $y = a^\bullet \Rightarrow y' = (a^\bullet \cdot \ln a) \times \bullet'$

(Ex) $y = 3^{5x^3 + 7x} \Rightarrow y' = (3x^{5x^3 + 7x} \times \ln 3) \cdot (15x^2 + 7)$

ⓔ $y = \log_a\bullet \Rightarrow y' = \dfrac{1}{\bullet \cdot \ln a} \times \bullet'$, $y = \ln\bullet \Rightarrow y' = \dfrac{1}{\bullet} \times \bullet'$

(Ex) $y = \log_3(2x^5) \Rightarrow y' = \dfrac{1}{2x^5 \times \ln 3} \times 10x^4$

ⓕ $y = f(\bullet) \Rightarrow y' = f'(\bullet) \cdot \bullet'$

(Ex) $y = f(10x) \Rightarrow y' = 10 \cdot f'(10x)$

반드시 알아 두어야 할 사항

① $f(x) = e^{\ln g(x)} = g(x)^{\ln e} = g(x)$

② $f(x) = \sqrt[m]{(\ )^n} = (\ )^{\frac{n}{m}}$ 으로 바꾸어서 $f'(x)$를 구하자.

③ $f(x) = \dfrac{a}{(\ )^n} = a(\ )^{-n}$ 으로 바꾸어서 $f'(x)$를 구하자.

④ $\sin^n f(x)$은 $\sin(f(x))^n$으로 놓고 $\dfrac{dx}{dy}$를 구하자.

⑤ $y = (\text{Constant})^{(\text{variable})} \Rightarrow$ 공식대로! (EX) $y = 2^x$

⑥ $y = (\text{Variable})^{(Variable)} \Rightarrow$ 양변에 ln을 취한다. (EX) $y = x^x$

⑦ Exponent가 애매한 경우에는 양변에 ln을 취한다.

⑧ $\dfrac{d}{dx}((\text{Variable})^{(Variable)})$에서 $y = (\text{Variable})^{(Variable)}$로 놓고 양변에 ln을 취한 후 $y' = \dfrac{dy}{dx}$ 임을 이용한다.

⑨ $y = (\text{Variable})^{(Constant)} \Rightarrow$ 공식대로! (EX) $y = x^2$

3. $\dfrac{dy}{dx}$, $\dfrac{d^2y}{dx^2}$

I. $\dfrac{dy}{dx}$

다음을 보자.

ⓐ $\dfrac{d}{dx}(x^2) = 2x$, ⓑ $\dfrac{d}{d\theta}(2\theta) = 2$, ⓒ $\dfrac{d}{dy}(y) = 1$, ⓓ $\dfrac{d}{dr}(r^2+1) = 2r$ …에서 보는 바와 같이

분모(Denominator)의 문자와 같은 식만 미분(Differentiation)이 가능함을 알 수 있다.

$\dfrac{d}{dx}(y)$ … 분모의 문자와 미분하려는 식이 문자가 다르다. 이럴 때는 $\dfrac{d}{dy}(y)\dfrac{dy}{dx} = \dfrac{dy}{dx}$가 된다.

그럼, 다음의 예제들을 보자.

ⓐ $\dfrac{d}{dx}(y^2) = \dfrac{d}{dy}y^2\dfrac{dy}{dx} = 2y\dfrac{dy}{dx}$

ⓑ $\dfrac{d}{dx}(3r) = \dfrac{d}{dr}(3r)\dfrac{dr}{dx} = 3\dfrac{dr}{dx}$

ⓒ $\dfrac{d}{dx}(3\theta^5) = \dfrac{d}{d\theta}(3\theta^5)\dfrac{d\theta}{dx} = 15\theta^4\dfrac{d\theta}{dx}$ …

위의 ⓐ, ⓑ, ⓒ의 경우에는 간단히 다음처럼 생각해서 풀어도 된다.

분모(Denominator)와 문자가 다를 때는 일단 그냥 미분 (Differentiation)을 하고 미분한 것에

$\dfrac{dy}{dx}$, $\dfrac{dr}{dx}$, $\dfrac{d\theta}{dx}$ **등을 취한다.**

다음을 보자.

ⓐ $y = x^2$ 에서 y를 미분하면 1이고 x가 아닌 것을 미분했으므로 미분한 것에 $\dfrac{dy}{dx}$를 취한다.

그러므로, $\dfrac{dy}{dx} = 2x$.

ⓑ $r = 3\sin\theta$ 에서 $\dfrac{dr}{d\theta}$을 구하려면 r을 미분하면 1이고 미분한 것에 $\dfrac{dr}{d\theta}$을 취하면 $\dfrac{dr}{d\theta} = 3\cos\theta$

가 된다. 이와 같이 알아 두는 것이 편할 때가 많다.

II. $\dfrac{d^2y}{dx^2}$, … , $\dfrac{d^ny}{dx^n}$

여기에서 $\dfrac{d^2y}{dx^2} = \dfrac{d}{dx}\dfrac{dy}{dx}$임을 알아 두자. 또, $\dfrac{d^2y}{dx^2} = f''(x) = y''$임을 알아 두자.

III. $\dfrac{d^2y}{dx^2} = \dfrac{d}{dx}\dfrac{dy}{dx}$

⇒ ① $\dfrac{dy}{dx}$를 먼저 구한 다음

② $\dfrac{d}{dx}$를 구한다.

19. If $x^2 + x^3 y = 1$, then when $x = 1$, $\dfrac{dy}{dx}$ is

(A) -4 (B) -2 (C) 0 (D) 1

20. If $y = \arcsin(\cos x)$ and x is an acute angle, then $\dfrac{dy}{dx}$ is

(A) $-\tan x$ (B) $\tan x$ (C) $\cot x$ (D) -1

21. If $\dfrac{dy}{dt} = \sqrt{5 - y^2}$, then $\dfrac{d^2 y}{dt^2}$ is

(A) 1 (B) $-y$ (C) $\dfrac{1}{\sqrt{5 - y^2}}$ (D) $\sqrt{5 - y^2}$

22. If $f(x) = x^2 + 2x$ and $g(x) = \sin x$, then $(g \circ f)'$ is

(A) $-\sin(x^2 + 2x)$

(B) $-2(x + 1) \cdot \cos(2x + 2)$

(C) $2(x + 1) \cdot \cos(x^2 + 2x)$

(D) $\cos(x^2 + 2x)$

23. If $\sin(xy) = \ln(x)$, then $\dfrac{dy}{dx}$ is

(A) $\dfrac{y - xy\sin(xy)}{x \cdot \cos(xy)}$

(B) $\dfrac{xy - \cos(xy)}{x^2\cos(xy)}$

(C) $1 - \dfrac{y}{x \cdot \cos(xy)}$

(D) $\dfrac{1 - xy \cdot \cos(xy)}{x^2 \cdot \cos(xy)}$

x	$f(x)$	$f'(x)$	$g(x)$	$g'(x)$
1	-2	5	-3	1
2	3	2	-2	-5
3	1	4	2	2

24. The table above gives values of f, f', g, and g' at selected value of x. If $h(x) = (f \circ g)(x)$, then $h'(3)$ is

(A) -4 (B) 1   (C) 4   (D) 8

25. If $x^2 + y^2 = 5$, what is the value of $\dfrac{d^2y}{dx^2}$ at the point (2, 1)?

(A) -5 (B) -3 (C) 0   (D) $\dfrac{1}{4}$

Solution

19. (B)

$x = 1$이면 $y = 0$

$2xx' + 3x^2 x' y + x^3 y' = 1'$

$\Rightarrow 2x + 3x^2 y + x^3 \dfrac{dy}{dx} = 0$

$\Rightarrow 2 + \dfrac{dy}{dx} = 0$

$\therefore \dfrac{dy}{dx} = -2$

20. (D)

$y = \sin^{-1}(\cos x) \Rightarrow y' = \dfrac{1}{\sqrt{1 - \cos^2 x}} \cdot (\cos x)'$

$\Rightarrow y' = \dfrac{-\sin x}{\sqrt{\sin^2 x}} = \dfrac{-\sin x}{\sin x} = -1$

21. (B)

$\dfrac{d^2 y}{dt^2} = \dfrac{d}{dt}\dfrac{dy}{dt} = \dfrac{d}{dt}(5 - y^2)^{\frac{1}{2}} = \dfrac{dy}{dt}\dfrac{d}{dy}(5 - y^2)^{\frac{1}{2}}$

$= \dfrac{dy}{dt}\left(\dfrac{1}{2}(5 - y^2)^{-\frac{1}{2}} \cdot (-2y)\right) = \dfrac{dy}{dt}\left(\dfrac{-y}{\sqrt{5 - y^2}}\right)$

$= \sqrt{5 - y^2}\left(\dfrac{-y}{\sqrt{5 - y^2}}\right) = -y$

22. (C)

$(g \circ f)' = (g(f(x))' = g'(f(x)) \cdot f'(x)$

$f(x) = x^2 + 2x, \ f'(x) = 2x + 2, \ g(x) = \sin x, \ g'(x) = \cos x$

그러므로, $\cos(x^2 + 2x) \cdot (2x + 2)$

23. (D)

$\cos(xy) \cdot (xy)' = \dfrac{1}{x}$ 에서 $\cos(xy) \cdot (x'y + xy') = \dfrac{1}{x}$

$\Rightarrow \cos(xy) \cdot (y + x\dfrac{dy}{dx}) = \dfrac{1}{x}$

$\Rightarrow y + x\dfrac{dy}{dx} = \dfrac{1}{x \cdot \cos(xy)}$

$\Rightarrow x\dfrac{dy}{dx} = \dfrac{1 - yx \cdot \cos(xy)}{x \cdot \cos(xy)}$

$\Rightarrow \dfrac{dy}{dx} = \dfrac{1 - xy \cdot \cos(xy)}{x^2 \cos(xy)}$

24. (C)

$h(x) = f(g(x)) \Rightarrow h'(x) = f'(g(x)) \cdot g'(x)$
$\Rightarrow h'(3) = f'(g(3)) \cdot g'(3) = f'(2) \cdot g'(3) = 2 \cdot 2 = 4$

25. (A)

$\dfrac{d^2y}{dx^2} = \dfrac{d}{dx}\dfrac{dy}{dx}$ 에서 $\dfrac{dy}{dx}$ 를 먼저 구한다.

$2xx' + 2yy' = 5' \Rightarrow 2x + 2y\dfrac{dy}{dx} = 0 \Rightarrow \dfrac{dy}{dx} = -\dfrac{x}{y}$

$\Rightarrow \dfrac{d}{dx}\dfrac{dy}{dx} = -\dfrac{d}{dx}(\dfrac{x}{y}) = -\dfrac{x'y - xy'}{y^2}$

$= -\dfrac{y - x(-\dfrac{x}{y})}{y^2} = -\dfrac{y + \dfrac{x^2}{y}}{y^2} = -\dfrac{1 + 2^2}{1} = -5$

Topic 7 - The derivative of inverse function

$$(f^{-1})'(y) = \frac{1}{f'(x)}$$

$y = f(x)$에서 y로부터 x값을 찾는다

x	$f(x)$	$f'(x)$
1	8	-2
2	5	-1
3	0	-4

26. The table above gives selected values for a differentiable function f and its derivative. If g is the inverse function of f, what is the value of $g'(5)$?

(A) -4 (B) -2 (C) -1 (D) 1

27. Let f be the function defined by $f(x)=x^2+x+\ln x$. If $h(x)=f^{-1}(x)$ for all x and the point $(1,2)$ is on the graph of f, what is the value of $h'(2)$?

(A) $\dfrac{1}{4}$ (B) $\dfrac{1}{2}$ (C) 1 (D) 4

Solution

26. (C)

$$g'(5) = (f^{-1})'(5) = \frac{1}{f'(2)} = -1$$

27. (A)

$$h'(2) = (f^{-1})'(2) = \frac{1}{f'(1)} \text{에서 } f'(x) = 2x + 1 + \frac{1}{x}$$

$$\Rightarrow f'(1) = 2 + 1 + 1 = 4$$

그러므로, $h'(2) = \dfrac{1}{f'(1)} = \dfrac{1}{4}$

Topic 8 - Integral 계산

앞에서 "적분(Integral)은 미분(Differentiation)의 반대"라고 공부했다. 여기에서부터 기본적인 공식이 시작된다.

$$\frac{1}{2}x^2 + C \xleftarrow[\text{적분(integral)}]{\text{미분(Differentiation)}} x$$

즉, x를 적분하면 $\frac{1}{2}x^2 + C$ 가 된다.

$\Rightarrow \int x^1 dx \Rightarrow \frac{1}{1+1}x^{1+1} + C \Rightarrow$ 이를 공식화 시키면…

반드시 암기하자!

$$\int x^n dx = \frac{1}{n+1}x^{n+1} + C$$

Integral은 다음과 같은 성질들도 있다. 반드시 알아 두자.

반드시 암기하자!

$$\int (f(x) \pm g(x))dx = \int f(x)dx \pm \int g(x)dx$$

$$\int af(x)dx = a\int f(x)dx$$

다음의 간단한 예제들을 통해서 위의 공식들을 확실하게 해 두자.

① $\int (x^2 - 2x + 3)dx = \int x^2 dx - 2\int x dx + \int 3dx = \frac{1}{3}x^3 - 2 \cdot \frac{1}{2}x^2 + 3x + C$

$(\int 3dx = \int 3x^0 dx = 3 \times \frac{1}{0+1}x^{0+1} + C = 3x + C \quad)$

② $\int (x^5 - 2x^4 + 7)dx = \frac{1}{6}x^6 - \frac{2}{5}x^5 + 7x + C$

③ $\int (t^3 - 2t)dt = \frac{1}{4}t^4 - t^2 + C$

Integration Formulas

① $\displaystyle\int \sin x\, dx = -\cos x + C$ $\qquad$ ($\cos x \xrightleftharpoons[\text{Integration}]{\text{Differentiation}} -\sin x$ or $\cos x \rightleftarrows -\sin x$)

② $\displaystyle\int \cos x\, dx = \sin x + C$ $\qquad$ ($\sin x \xrightleftharpoons[\text{Integration}]{\text{Differentiation}} \cos x$ or $\sin x \rightleftarrows \cos x$)

③ $\displaystyle\int \sec^2 x\, dx = \tan x + C$ $\qquad$ ($\tan x \xrightleftharpoons[\text{Integration}]{\text{Differentiation}} \sec^2 x$ or $\tan x \rightleftarrows \sec^2 x$)

④ $\displaystyle\int \csc^2 x\, dx = -\cot x + C$ $\qquad$ ($\cot x \xrightleftharpoons[\text{Integration}]{\text{Differentiation}} -\csc^2 x$ or $\cot x \rightleftarrows -\csc^2 x$)

⑤ $\displaystyle\int \sec x \cdot \tan x\, dx = \sec x + C$ $\qquad$ ($\sec x \xrightleftharpoons[\text{Integration}]{\text{Differentiation}} \sec x \tan x$ or

$\sec x x \rightleftarrows \sec x \tan x$)

⑥ $\displaystyle\int \csc x \cdot \cot x\, dx = -\csc x + C$ $\qquad$ ($\csc x \xrightleftharpoons[\text{Integration}]{\text{Differentiation}} -\csc x \tan x$ or

$\csc x \rightleftarrows -\csc x \cot x$)

⑦ $\displaystyle\int \frac{1}{x}\, dx = \ln|x| + C$ $\qquad$ ($\ln x \xrightleftharpoons[\text{Integration}]{\text{Differentiation}} \frac{1}{x}$ or $\ln x \rightleftarrows \frac{1}{x}$)

⑧ $\displaystyle\int e^x\, dx = e^x + C$ $\qquad$ ($e^x \xrightleftharpoons[\text{Integration}]{\text{Differentiation}} e^x$ or $e^x \rightleftarrows e^x$)

⑨ $\displaystyle\int a^x\, dx = \frac{a^x}{\ln a} + C$ $\qquad$ ($a^x \xrightleftharpoons[\text{Integration}]{\text{Differentiation}} a^x \ln a$ or $a^x \rightleftarrows a^x \ln a$)

⑩ $\displaystyle\int \frac{1}{\sqrt{1-x^2}}\, dx = \sin^{-1} x + C$

$\qquad$ ($\sin^{-1} x \xrightleftharpoons[\text{Integration}]{\text{Differentiation}} \frac{1}{\sqrt{1-x^2}},\ -1 < x < 1$ or $\sin^{-1} x \rightleftarrows \frac{1}{\sqrt{1-x^2}},\ -1 < x < 1$)

⑪ $\displaystyle\int \frac{1}{1+x^2}\, dx = \tan^{-1} x + C$

$\qquad$ ($\tan^{-1} x \xrightleftharpoons[\text{Integration}]{\text{Differentiation}} \frac{1}{1+x^2}$ or $\tan^{-1} x \rightleftarrows \frac{1}{1+x^2}$)

⑫ $\displaystyle\int \frac{1}{|x|\sqrt{x^2-1}}\, dx = \sec^{-1} x + C$

$\qquad$ ($\sec^{-1} x \xrightleftharpoons[\text{Integration}]{\text{Differentiation}} \frac{1}{|x|\sqrt{x^2-1}},\ |x| > 1$ or $\sec^{-1} x \rightleftarrows \frac{1}{|x|\sqrt{x^2-1}},\ |x| > 1$)

⑬ $\displaystyle\int \sec x\, dx = \ln|\sec x + \tan x| + C$

⑭ $\displaystyle\int \csc x\, dx = \ln|\csc x - \cot x| + C$

U-Substitution

Integral을 계산하는 데 있어서 가장 중요한 부분이다. 많은 학생들이 Integral을 계산할 때 사소한 부분을 자주 실수하는데 이는 무턱대고 공식 따라 해결하려고 하기 때문에 그런 일이 발생한다.

조금 귀찮더라도 다음에 소개하는 7가지가 눈에 보인다면 바로 "치환(Substitution)"을 하여야 한다.

〈치환을 해야 하는 7가지 규칙〉

① $\displaystyle\int \bullet \times \bigcirc \, dx$

(Example) $\displaystyle\int (x^3 + 2x - 3)(6x^2 + 4) dx$

(Solution)

$x^3 + 2x - 3$ 을 미분하면 $3x^2 + 2$ 이므로 $6x^2 + 4 = 2(3x^2 + 2)$ 와 미분 결과가 같다.

그러므로 $x^3 + 2x - 3 = u$ 라고 치환하고 양변을 x 에 대해서 미분하면 $3x^2 + 2 = \dfrac{du}{dx}$.

그러므로, $dx = \dfrac{1}{3x^2 + 2} du$. 그러므로, $\displaystyle\int 2u(3x^2 + 2)\dfrac{1}{3x^2 + 2} du = 2\int u \, du = u^2 + C$,

$u = x^3 + 2x - 3$ 이므로 결과는 $(x^3 + 2x - 3)^2 + C$

---

② $\displaystyle\int \dfrac{\bigcirc}{\bullet} \, dx$

(Example) $\displaystyle\int \dfrac{t-1}{\sqrt{t^2 - 2t}} dt$

(Solution)

$t^2 - 2t$ 을 미분하면 $2t - 2 = 2(t - 1)$ 이므로 $t^2 - 2t = u$ 라고 치환하고 양변을 t 에 대해서 미분하면 $2t - 2 = \dfrac{du}{dt}$ 에서 $dt = \dfrac{1}{2(t-1)} du$ 이다.

그러므로, $\displaystyle\int \dfrac{t-1}{\sqrt{u}} \cdot \dfrac{1}{2(t-1)} du = \dfrac{1}{2}\int u^{-\frac{1}{2}} du = \dfrac{1}{2} \times \dfrac{1}{1 - \frac{1}{2}} u^{-\frac{1}{2}+1} + C = u^{\frac{1}{2}} + C$

$u = t^2 - 2t$ 이므로 결과는 $\sqrt{t^2 - 2t} + C$

③ $\int \bigcirc \times a^{\bullet} dx$

(Example) $\int \sin x \cdot e^{\cos x} dx$

(Solution)

$\cos x$를 미분하면 $-\sin x$ 이므로 $\cos x = u$ 라고 치환하고 양변을 x에 대해서 미분하면

$-\sin x = \dfrac{du}{dx}$에서 $dx = -\dfrac{1}{\sin x} du$ 이다. 그러므로,

$\int \sin x \times e^u \times (-\dfrac{1}{\sin x}) du = -\int e^u du = -e^u + C$ 이고 $u = \cos x$ 이므로 결과는 $-e^{\cos x} + C$

④ $\int \bigcirc \cdot \sin \bullet \, dx$

(Example) $\int (x-3) \cdot \sin(x^2 - 6x) dx$

(Solution)

$x^2 - 6x$를 미분하면 $2x - 6 = 2(x-3)$이므로 $x^2 - 6x = u$라고 치환하고 양변을 x에 대해서 미분하면 $2x - 6 = \dfrac{du}{dx}$에서 $dx = \dfrac{1}{2(x-3)} du$이다.

그러므로, $\int (x-3) \times \sin u \times \dfrac{1}{2(x-3)} du = \dfrac{1}{2} \int \sin u \, du = -\dfrac{1}{2} \cos u + C$ 이고

$u = x^2 - 6x$ 이므로 결과는 $-\dfrac{1}{2} \cos(x^2 - 6x) + C$

⑤ $\int \dfrac{\bigcirc}{1 + \bullet^2} dx, \quad \int \dfrac{\bigcirc}{1 - \bullet^2} dx$

(Example) $\int \dfrac{15x^2}{1 + (5x^3)^2} dx$

(Solution)

$5x^3$를 미분하면 $15x^2$이므로 $5x^3 = u$ 라고 치환하고 양변을 x에 대해서 미분하면

$15x^2 = \dfrac{du}{dx}$에서 $dx = \dfrac{1}{15x^2} du$ 이다.

그러므로, $\int \dfrac{15x^2}{1 + u^2} \cdot \dfrac{1}{15x^2} du = \tan^{-1} u + C$ 이므로 결과는 $\tan^{-1} 5x^3 + C$

⑥ $\int (일차식)^n dx$

(Example) $\int (7x-5)^5 dx$

(Solution)

$7x-5=u$ 라고 치환하고 양변을 x에 대해서 미분하면 $7=\dfrac{du}{dx}$에서 $dx=\dfrac{1}{7}du$이므로

$\dfrac{1}{7}\int u^5 du = \dfrac{1}{7}\times\dfrac{1}{6}u^6+C=\dfrac{1}{42}u^6+C$, $u=7x-5$ 이므로 결과는 $\dfrac{1}{42}(7x-5)^6+C$

⑦ ax꼴은 치환!

(Example) $\int \dfrac{3}{\sqrt{1-(5x)^2}}dx$

(Solution)

$5x=u$라고 치환하고 양변을 x에 대해서 미분하면 $5=\dfrac{du}{dx}$에서 $dx=\dfrac{1}{5}du$이므로

$\int \dfrac{3}{\sqrt{1-u^2}} \cdot \dfrac{1}{5}du = \dfrac{3}{5}\int \dfrac{1}{\sqrt{1-u^2}}du = \dfrac{3}{5}\sin^{-1}u+C$ 이고 $u=5x$이므로

결과는 $\dfrac{3}{5}\sin^{-1}5x+C$

"U-Substitution"의 Integration 문제를 푸는 데 있어서 많은 학생들이 다음을 빨리 찾아내지 못하는 것 같아서 다음과 같이 정리해보았다.

<u>잘 보이지 않는 Derivative 결과</u>

$\sin^{-1}x \Rightarrow$ (Derivative) $\Rightarrow \dfrac{1}{\sqrt{1-x^2}}$

$\tan^{-1}x \Rightarrow$ (Derivative) $\Rightarrow \dfrac{1}{1+x^2}$

$\sin^2 x \Rightarrow$ (Derivative) $\Rightarrow \sin 2x$

$\cos^2 x \Rightarrow$ (Derivative) $\Rightarrow -\sin 2x$

$\sqrt{x} \Rightarrow$ (Derivative) $\Rightarrow \dfrac{1}{2\sqrt{x}}$

$\dfrac{1}{x} \Rightarrow$ (Derivative) $\Rightarrow -\dfrac{1}{x^2}$

$\ln x \Rightarrow$ (Derivative) $\Rightarrow \dfrac{1}{x}$

뻔히 알고 있는 결과인데도 Integration 문제를 풀다가 보면 특히 위의 다섯 가지가 눈에 확 안 들어올 때가 많다.

Definite Integrals

"Indefinite Integrals"와는 달리 "Definite Integrals"에서는 범위가 주어지게 되므로 다음과 같은 규칙들이 나오게 된다. 읽어보면 너무 뻔한 이야기 같지만 중요한 것은 **이 규칙들을 모두 암기!** 해야 한다는 사실~!!

반드시 암기하자!

① $\displaystyle\int_a^b f(x)dx = F(b) - F(a)$

② $\displaystyle\int_a^a f(x)dx = 0$

③ $\displaystyle\int_a^b f(x)dx = \int_a^c f(x)dx + \int_c^b f(x)dx$: 모두 $f(x)$로 같을 때

④ $\displaystyle\int_a^b f(x)dx = -\int_b^a f(x)dx$

28. $\int_{-1}^{1} \frac{|x|}{x} dx =$

(A) -1 (B) 0 (C) 1 (D) $\frac{1}{2}$

29. $\int \left(\frac{1}{x} \int_{1}^{x} \frac{1}{t} dt \right) dx =$

(A) $\frac{1}{2}(\ln x)^2 + C$

(B) $(\ln x)^2 + C$

(C) $\frac{1}{2}\ln x + C$

(D) $\frac{1}{x}$

30. $\int \frac{1}{\sqrt{16-t^2}} dt =$

(A) $\frac{1}{4}\sin^{-1}\left(\frac{t}{4}\right) + C$

(B) $\frac{1}{4}\sin^{-1} t + C$

(C) $\sin^{-1} t + C$

(D) $\sin^{-1}\left(\frac{t}{4}\right) + C$

31. $\int_{0}^{\frac{\pi}{3}} \sec^2 x\, e^{\tan x}\, dx =$

(A) $e^{\sqrt{3}}$

(B) $e^{\sqrt{3}} - 1$

(C) $e^{\sqrt{3}} + 1$

(D) $e^3 + C$

32. If f is a continuous function and if $F'(x) = f(x)$ for all real numbers x, then $\int_{2}^{4} f(5x)\, dx =$

(A) $\dfrac{1}{5} F(2) - \dfrac{1}{5} F(0)$

(B) $F(4) - F(2)$

(C) $\dfrac{1}{5} F(4) - \dfrac{1}{5} F(2)$

(D) $\dfrac{1}{5} F(20) - \dfrac{1}{5} F(10)$

33. $\dfrac{d}{dx} \int_{0}^{x^2} \cos(t)\, dt =$

(A) $2x^2 \cos(x)$

(B) $x \cos(x)$

(C) $2x \cos(x)$

(D) $2x \cos(x^2)$

34. $\int \dfrac{x^2}{x^2+1}\,dx =$

(A) $x - \tan^{-1}x + C$

(B) $x^2 + C$

(C) $\tan^{-1}(1+x^2)+ C$

(D) $\ln|x^2+1|+ \dfrac{1}{3}x^3 + C$

35. If $\displaystyle\int_0^1 f(x)\,dx = 3$ and $\displaystyle\int_4^0 f(x)\,dx = 5$, then $\displaystyle\int_1^4 (2f(x)+2)\,dx =$

(A) -10 (B) -8 (C) 2 (D) 4

⚠️ Solution

28. (B)

$$\int_{-1}^{1} \frac{|x|}{x} dx = \int_{-1}^{0} (-1) dx + \int_{0}^{1} (1) dx = [-x]_{-1}^{0} + [x]_{0}^{1}$$

$$= (0-1) + (1-0) = 0$$

29. (A)

① $\int_{1}^{x} \frac{1}{t} dt = [\ln t]_{1}^{x} = \ln x$

② $\int \frac{1}{x} \cdot \ln x\, dx \implies \ln x = u, \ \frac{1}{x} = \frac{du}{dx}$

$\implies \int \frac{1}{x} \cdot u \cdot x\, du = \int u\, du = \frac{1}{2} u^2 + C$

∴ $\frac{1}{2} (\ln x)^2 + C$

30. (D)

$$\int \frac{1}{\sqrt{16(1 - (\frac{1}{4}t)^2)}} dt \text{에서} \ \frac{1}{4}t = u, \frac{1}{4} = \frac{du}{dt}$$

$$\implies \frac{1}{4} \int \frac{1}{\sqrt{1 - u^2}} \cdot 4du = \int \frac{1}{\sqrt{1 - u^2}} du = \sin^{-1} u + C$$

∴ $\sin^{-1}(\frac{1}{4}t) + C$

31. (B)

$$\tan x = u, \ \sec^2 x = \frac{du}{dx}$$

$$\int_{0}^{\sqrt{3}} \sec^2 x \cdot e^u \frac{du}{\sec^2 x} \implies \int_{0}^{\sqrt{3}} e^u du$$

$$\implies [e^u]_{0}^{\sqrt{3}} = e^{\sqrt{3}} - 1$$

32. (D)

$5x = u, \ 5 = \dfrac{du}{dx}$

$\dfrac{1}{5} \displaystyle\int_{10}^{20} f(u)\,du = \dfrac{1}{5}(F(20) - F(10)) = \dfrac{1}{5}F(20) - \dfrac{1}{5}F(10)$

33. (D)

$\cos(t) = f(t)$라고 하면,

$\displaystyle\int_{0}^{x^2} f(t)\,dt = F(x^2) - F(0)$

$\dfrac{d}{dx}(F(x^2) - F(0)) = 2xf(x^2) = 2x \cdot \cos x^2$

34. (A)

$\displaystyle\int \dfrac{x^2}{x^2 + 1}\,dx = \int \dfrac{x^2 + 1 - 1}{x^2 + 1}\,dx = \int \left(1 - \dfrac{1}{1 + x^2}\right)dx = x - \tan^{-1}x + C$

35. (A)

$\displaystyle\int_{0}^{1} f(x)\,dx + \int_{4}^{0} f(x)\,dx = \int_{4}^{1} f(x)\,dx = 8 \ \Rightarrow \ \int_{1}^{4} f(x)\,dx = -8$

$\Rightarrow \displaystyle\int_{1}^{4}(2f(x) + 2)\,dx = 2\int_{1}^{4} f(x)\,dx + \int_{1}^{4} 2\,dx$

$= 2(-8) + [2x]_{1}^{4} = -16 + (8 - 2) = -10$

Topic 9 - Graph

AP Calculus AB & BC 시험에서는 Graph의 유추만 출제가 되고 그리는 것은 거의 출제가 되지 않는다. 대부분의 Graph는 정확하게 그리기가 어렵다. 다음에서 소개하는 Graph들 또한 정확하게 그린다기보다는 어느 정도 비슷하게 추정을 하는 것이다.

Graph의 추정 방법

① f Graph가 Increasing $\Rightarrow$ f'은 Positive
② f Graph가 Decreasing $\Rightarrow$ f'은 Negative

f가 differentiable하므로 f'이 존재한다.
그러나 f'은 continuous하지 않을 수 있다.
이해를 돕고자 이와 같이 좀 오버해서 그렸지만
실제로 이와 같은 그림은 거의 그려지지 않는다.

※ 실제로, ②와 같은 그래프는 그려지지 않는다. 여기서 말하고자 하는 것은 "함수 f가 $x = a$에서 differentiable 하다고 해서 f'이 $x = a$에서 항상 연속임을 의미하지 않는다는 것이다."

그러면, 언제 "함수 f가 $x = a$에서 differentiable 하지만 f'이 $x = a$에서 continuous 하지 않을까?" 다음의 예를 살펴보자.

$$f(x) = \begin{cases} x^2\sin\dfrac{1}{x} & (x \neq 0) \\ 0 & (x = 0) \end{cases}$$

$\Rightarrow$ ① $\lim\limits_{x \to 0} f(x) = f(0)$ 이므로 f는 $x = 0$에서 continuous 하다.

$\Rightarrow$ ② $f'(x) = 2x\sin\dfrac{1}{x} - (\cos\dfrac{1}{x})$. $\lim\limits_{x \to 0+} f'(x)$ and $\lim\limits_{x \to 0-} f'(x)$는 진동하므로 $\lim\limits_{x \to 0} f'(x) = f'(0)$인지 확신할 수 없다. 일반적인 것은 아니지만, 이와 같은 사례가 있음을 이해하고 있어야 한다.

- f'이 cusp인 것이지 f가 cusp가 아니다.
- f'값이 Exist하므로 Differentiable

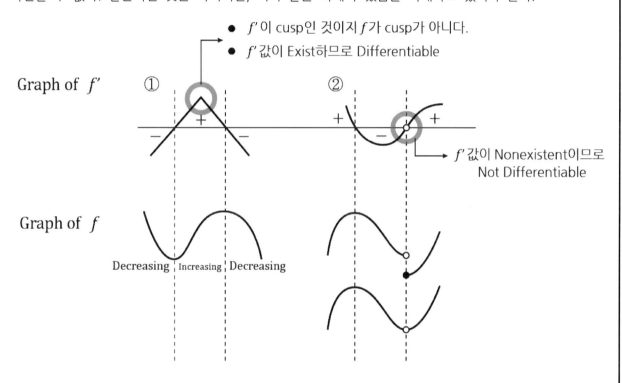

Graph of f'

Graph of f

②번의 그림에서와 같이 f'값이 Nonexistent하여 "Not Differentiable"일 때에는 그림과 같이 f의 Graph를 여러 가지로 생각해 볼 수 있다.

Polynomial Function Graph 그리기

I. Polynomial function이란?

$y = ax^n + bx^{n-1} + cx^{n-2} + \cdots + z \cdots$ 와 같이 생긴 Function $\cdots$
Polynomial function은 그 모양이 어느 정도 정해져 있다.

다음과 같이 알아 두자.
$y = ax^n + bx^{n-1} + cx^{n-2} + \cdots + z \cdots$ 에서

	$a > 0$ (오른쪽 끝이 위로)	$a < 0$ (오른쪽 끝이 아래로)
① $y = ax$	$a > 0$	$a < 0$
② $y = ax^2 + bx + c$	$a > 0$	$a < 0$
③ $y = ax^3 + bx^2 + cx + d$	$a > 0$	$a < 0$
④ $y = ax^4 + bx^3 + cx^2 + dx + e$	$a > 0$	$a < 0$

II. Polynomial Function Graph 그리기

예를 들어, $f(x) = (x-1)(x-2)^2(x-3)^3(x-4)^4$의 개형을 그려서 부분적으로 자세히 보면 다음과 같다.

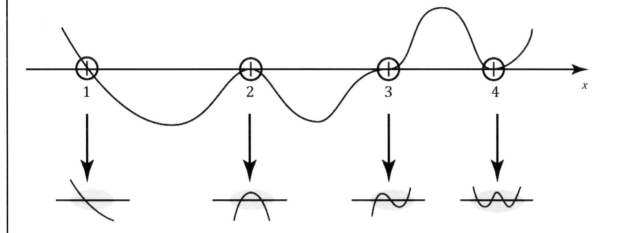

위의 그림에서 보는 것처럼 $f(x) = (\quad)^n$에서 n이 Even이면 x축에 접(tangent)하고 n이 Odd이면 x축을 지난다.

Graph 총정리

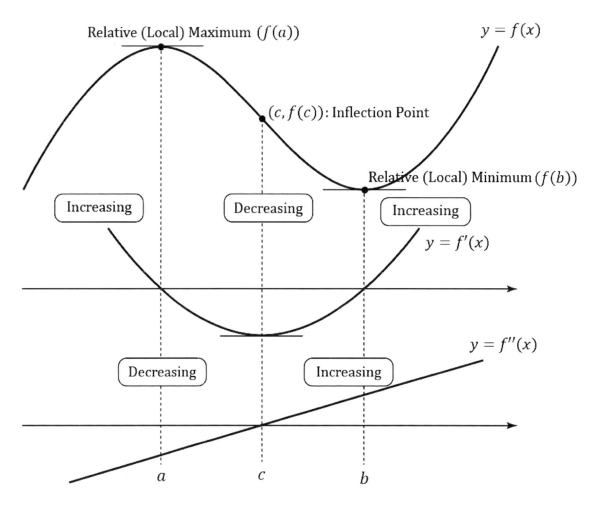

$y=f(x)$, $y=f'(x)$, $y=f''(x)$의 Graph를 보면서 알 수 있는 것들을 정리해보면, 다음과 같다. 너무나 중요한 내용이니 그림과 함께 숙달될 때까지 여러 번 써보면서 암기하도록 한다.

<u>다음의 내용은 너무나 중요하다!</u>

(1) Relative Maximum $f(a)$

① $f'(x)$가 Positive에서 Negative로 바뀌는 점.

② $f'(x)=0$ 이면서 $f''(x)<0$인 점.

(2) Relative Minimum $f(b)$

① $f'(x)$가 Negative에서 Positive로 바뀌는 점.

② $f'(x)=0$ 이면서 $f''(x)>0$인 점.

---

(3) Inflection Point $(c, f(c))$

- $y = f(x)$의 Graph의 Concavity가 바뀌는 점.
- The slope of the tangent line의 변화가 급격하게 일어나는 점.
① $f'(x)$가 Decreasing에서 Increasing으로 바뀌는 점.
 또는 Increasing에서 Decreasing으로 바뀌는 점.
② $f''(x)$의 부호가 바뀌는 점.

(4) Concave up

① $f'(x)$이 Increasing 하는 구간.
② $f''(x) > 0$인 구간, 즉, x값의 범위.

(5) Concave down

① $f'(x)$이 Decreasing 하는 구간.
② $f''(x) < 0$인 구간, 즉, x값의 범위.

(6) $y = f(x)$가 Increasing 하는 구간

① $f'(x) > 0$ 인 구간.

(7) $y = f(x)$가 Decreasing 하는 구간

① $f'(x) < 0$ 인 구간.

Exercise

36. If $f''(x) = (x-1)^2 \cdot x \cdot (x+1)^3$, then the graph of $f(x)$ has inflection point(s) when $x =$

(A) -1 and 0 (B) 0 and 1 (C) -1 and 1 (D) $-1, 0,$ and 1

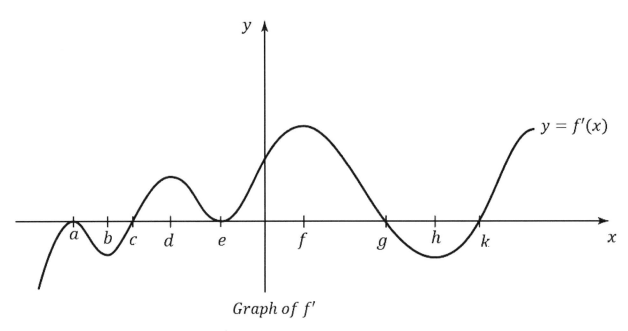

Graph of f'

37. The graph of f', the derivative of f is shown in the figure above. Which of the following must be true?

   (A) f has relative maxima at $x = c$ and k.
   (B) f has four inflection points.
   (C) f has relative minima at $x = e$ and g.
   (D) f has two relative minima and one relative maximum.

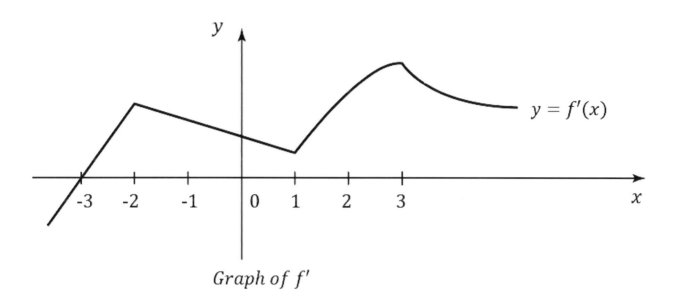

y

-3   -2   -1   0   1   2   3 x

$y = f'(x)$

Graph of f'

38. The graph of f', the derivative of f is shown in the figure above. Which of the following statements is true about f?

(A) f is not differentiable at $x = -2$, 1, and 3.
(B) f is decreasing for $-2 < x < 1$.
(C) f is both increasing and concave up for $1 < x < 3$.
(D) f has a local minimum at $x = 1$.

39. The function f has the property that $f(x) > 0$, $f'(x) > 0$, and $f''(x) > 0$ for all real values x. Which of the following could be the graph of f?

(A)

(B)

(C)

(D)

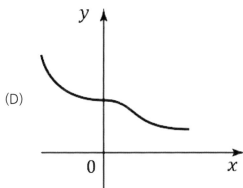

40. The function f has the property that $f'(x) < 0$ and $f''(x) > 0$ for all x in the closed interval $[1, 4]$. Which of the following could be a table of values for f?

(A)

x	$f(x)$
1	10
2	5
3	3
4	2

(B)

x	$f(x)$
1	10
2	9
3	7
4	2

(C)

x	$f(x)$
1	10
2	11
3	12
4	13

(D)

x	$f(x)$
1	2
2	7
3	9
4	10

Solution

36. (A)

f'' graph를 그려보면 대략 다음과 같다.

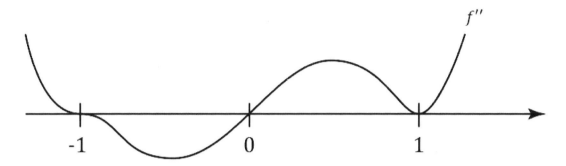

즉, f''의 sign change가 발생하는 점은 $x = -1, 0$이므로 f의 inflection point는 $x = -1$과 $x = 0$에서 발생한다.

37. (D)

• f' graph가 increasing에서 decreasing 또는 decreasing에서 increasing으로 바뀌는 점에서 f graph는 inflection point가 생기므로 주어진 f' graph로부터 f는 $x = a, b, d, e, f, h$에서 inflection point를 갖는다는 것을 알 수 있다.

• f' graph가 positive에서 negative로 바뀌는 점에서 f graph는 relative maximum이 생기므로 f는 $x = g$에서 relative maximum을 갖게 된다.

38. (C)

• f' graph가 increasing 하는 구간에서 f graph는 concave up이고 f' graph가 decreasing 하는 구간에서 f graph는 concave down이다.

$f' > 0$인 x의 구간에서 f는 increasing하고 $f' < 0$인 x의 구간에서 f는 decreasing 한다. 그러므로, $1 < x < 3$에서 f는 increasing 하면서 concave up이 된다. 정답은 (C).

(A) $x = -2, 1, 3$에서 f' 값이 존재하므로 f는 differentiable.

(B) $-2 < x < 1$에서 $f' > 0$이므로 f는 increasing 한다.

(D) f는 f'이 negative에서 positive로 바뀌는 점에서 local minimum을 가지므로 주어진 f' graph로부터 알 수 있는 것은 f는 $x = -3$일 때 local minimum을 갖는다는 것이다.

39. (B)

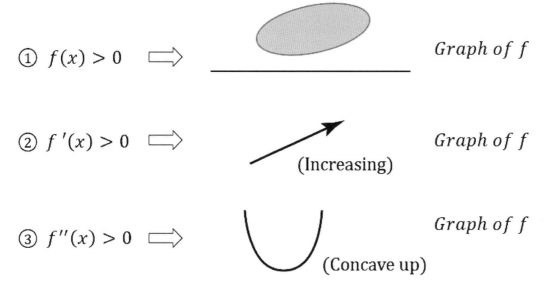

① $f(x) > 0$ ⟹   Graph of f

② $f'(x) > 0$ ⟹ (Increasing) Graph of f

③ $f''(x) > 0$ ⟹ (Concave up) Graph of f

그러므로, 위의 ①, ②, ③ 모두를 만족할 수 있는 Graph는 (B)번.

40. (A)
$f'(x) < 0$이고 $f''(x) > 0$이면 f Graph는 decreasing 하면서 concave up이다.
즉, f 는 다음과 같은 Graph가 된다.

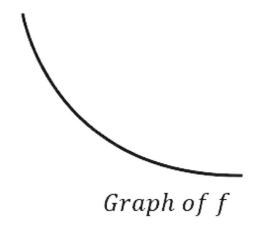

Graph of f

위와 같은 f graph 위에 있을 수 있는 값은 (A)번 Table이다.

Topic 10 - Related Rates

변화율을 구하는 단원이다. 비교적 문장이 긴 문제들을 풀어야 하는 경우가 많은데 알고 보면 생각보다 쉽다.

⇒ 변화율이란?
"짧은 시간 동안 일어나는 부피(Volume), 길이(Length), 반지름(Radius), ⋯ 등의 변화 비율"

다음과 같이 구한다.

① 구하고자 하는 목적이 무엇인지 파악한다.
(거의 문장 끝에 나온다. 길이의 변화율⋯, 부피의 변화율⋯)
② 구하고자 하는 주제에 대해서 식을 세운다.

(예를 들어, 원의 면적이면 $A = \pi r^2$, 구의 부피이면 $V = \dfrac{4}{3}\pi r^3$ ⋯ 등등 ⋯)

③ 양변을 시간 t에 대해서 미분(Derivative). 즉, 양변에 $\dfrac{d}{dt}$를 취한다.

④ 문장 중에 필요한 수치는 다 준다. 그대로 대입만 하면 된다.

Exercise

41. The radius of a circle is increasing at a constant rate of 2 meters per second. What is the rate of increase in the area of the circle at the instant when circumference of the circle is 100π meters?

(A) $100\pi\,\mathrm{m^2/sec}$ (B) $200\pi\,\mathrm{m^2/sec}$ (C) $400\pi\,\mathrm{m^2/sec}$ (D) $800\pi\,\mathrm{m^2/sec}$

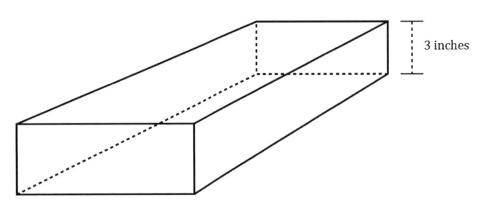

3 inches

42. In the rectangular prism above, the height is 3 inch. Its length increases at the rate of $2\,\mathrm{in/sec}$, its width increases at the rate of $3\,\mathrm{in/sec}$. When the length is 6 inch and the width is 12 inch, the rate, in cubic inches per second, at which the volume of the rectangular prism is changing is

(A) 48 (B) 96 (C) 126 (D) 154

📱 43. A man stands on the road 30 meters north of the crossing and watches an westbound car traveling at 15 meters per second. At hour many meters per second is the car moving away from the student 12 seconds after it passes through the intersection? (A road track and a road cross at right angles)

(A) 11.346   (B) 14.796   (C) 20.174   (D) 23.667

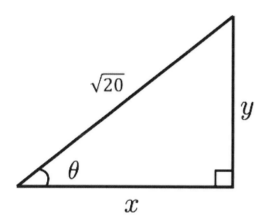

44. In the triangle shown above, if θ increases at a constant rate of 3 radians per second, at what rate is y increasing in units per second when y equals $\sqrt{13}$ units?

(A) $3\sqrt{7}$ (B) $\dfrac{\sqrt{7}}{\sqrt{20}}$ (C) $3\sqrt{13}$ (D) $\dfrac{\sqrt{13}}{\sqrt{20}}$

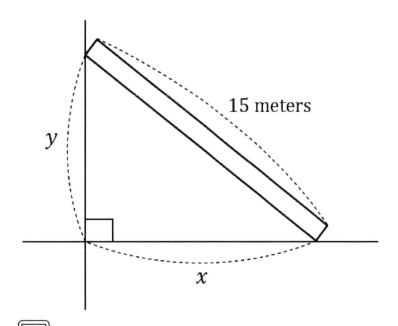

15 meters

y

x

45. In the figure above, a 15 meters long ladder is leaning against a wall and is sliding towards the floor. The very bottom of the ladder is sliding away from base of the wall at a rate of 5 m/sec. If the top of the ladder is vertically 4 meters away from the ground, what is the rate of change of the distance between the top of the ladder and the bottom?

(A) -20.245 m/sec

(B) -18.071 m/sec

(C) -15.349 m/sec

(D) -13.226 m/sec

Solution

41. (B)

Circle의 Area를 A라 하면,

① Topic : $\dfrac{dA}{dt}$

② Equation : $A = \pi r^2$

③ $\dfrac{d}{dt}$: $\dfrac{dA}{dt} = 2\pi r \dfrac{dr}{dt}$

④ • $100\pi = 2\pi r$에서 $r = 50$

  • $\dfrac{dr}{dt} = 2 \, \mathrm{m/sec}$

그러므로, $\dfrac{dA}{dt} = 2\pi \times 50 \times 2 = 200\pi \; \mathrm{m^2/sec}$

42. (C)

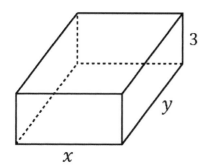

Rectangular Prism의 Volume을 V라고 하면,

① Topic : $\dfrac{dV}{dt}$

② Equation : $V = 3xy$

③ $\dfrac{d}{dt}$: $\dfrac{dV}{dt} = 3x'y + 3xy' = 3y\dfrac{dx}{dt} + 3x\dfrac{dy}{dt}$

④ $\dfrac{dx}{dt} = 2$, $\dfrac{dy}{dt} = 3$, $x = 6$, $y = 12$이므로 $\dfrac{dV}{dt} = 3 \times 12 \times 2 + 3 \times 6 \times 3 = 126$

43. (B)

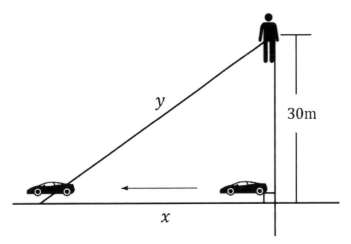

① Topic : $\dfrac{dy}{dt}$

② Equation : $y^2 = x^2 + 30^2$

③ $\dfrac{d}{dt}$: $2y\dfrac{dy}{dt} = 2x\dfrac{dx}{dt}$

④ $x = 15 \times 12 = 180$, $y^2 = 180^2 + 30^2$ $\Rightarrow$ $y \approx 182.483$

그러므로, $(2 \times 182.483)\dfrac{dy}{dt} = 2 \times 180 \times 15$에서 $\dfrac{dy}{dt} = 14.796$

44. (A)

① Topic : $\dfrac{dy}{dt}$

② Equation : $\sin\theta = \dfrac{y}{\sqrt{20}}$

③ $\dfrac{d}{dt}$: $(\cos\theta)\dfrac{d\theta}{dt} = \dfrac{1}{\sqrt{20}}\dfrac{dy}{dt}$

④ $\dfrac{d\theta}{dt} = 3$, $y = \sqrt{13}$, $x = \sqrt{7}$, $\cos\theta = \dfrac{\sqrt{7}}{\sqrt{20}}$

그러므로, $\dfrac{\sqrt{7}}{\sqrt{20}} \times 3 = \dfrac{1}{\sqrt{20}}\dfrac{dy}{dt}$에서 $\dfrac{dy}{dt} = 3\sqrt{7}$

45. (B)

① Topic : $\dfrac{dy}{dt}$

② Equation : $x^2 + y^2 = 15^2$

③ $\dfrac{d}{dt}$: $2x\dfrac{dx}{dt} + 2y\dfrac{dy}{dt} = 0$

④ $y = 4$, $\dfrac{dx}{dt} = 5$, $x^2 + y^2 = 15^2 \Rightarrow x^2 + 16 = 225 \Rightarrow x \approx 14.457$

그러므로, $2 \times 14.457 \times 5 + (2 \times 4)\dfrac{dy}{dt} = 0 \Rightarrow \dfrac{dy}{dt} \approx -18.071\,\mathrm{m/sec}$

Topic 11 - Motion

Position, $P(t)$ $\xrightarrow[\text{Integral}]{\text{Differentiation}}$ Velocity, $V(t)$ $\xrightarrow[\text{Integral}]{\text{Differentiation}}$ Acceleration, $A(t)$

$\quad\quad f(t)$ $\quad\quad\quad\quad\quad\quad\quad\quad\quad\quad f'(t)$ $\quad\quad\quad\quad\quad\quad\quad\quad\quad\quad f''(t)$

① Speed = | Velocity |

② Total Distance = $\displaystyle\int_a^b |\text{Velocity}|\, dt = \int_a^b (\text{Speed})\, dt$

③ $\begin{cases} P(b) - P(a) = \displaystyle\int_a^b V(t)\, dt \\ f(b) - f(a) = \displaystyle\int_a^b f'(t)\, dt \end{cases}$

$\quad \begin{cases} V(b) - V(a) = \displaystyle\int_a^b A(t)\, dt \\ f'(b) - f'(a) = \displaystyle\int_a^b f''(t)\, dt \end{cases}$

④ • Speed Increasing : Velocity와 Acceleration의 Sign이 같을 때
   • Speed Decreasing : Velocity와 Acceleration의 Sign이 다를 때

Exercise

46. The acceleration a of a particle moving in a straight line is given in terms of time t by $a(t) = 6 - 4t$. If the velocity of the particle is 4 at $t = 1$ and if $s(t)$ is the distance of the particle from the origin at time t, what is $s(2) - s(0)$?

(A) $\dfrac{20}{3}$ (B) $\dfrac{10}{3}$ (C) $\dfrac{7}{3}$ (D) 2

47. The velocity of a particle moving on the line of x−axis is given as $v(t) = e^t + t$. At time $t = 0$, the x−coordinate of this particle is $x(0) = 3$. If time $t = 2$, what is the x−coordinate of this particle?

(A) e^2 (B) $4 + e^2$ (C) $4 + 2e^2$ (D) $6 + 4e^2$

48. A particle moves along the x−axis so that at time $t \geq 0$, its position is given by $x(t) = 2t^3 - 21t^2 + 72t - 10$. What is the time interval at which the speed of this particle increases?

(A) $0 < t < 3$ (B) $3 < t < 3.5$ (C) $3.5 < t < 4$ (D) $3 < t < 4$

Solution

46. (A)

$v(t) = \int (6 - 4t)dt = 6t - 2t^2 + C$ 에서 $v(1) = 4 = 6 - 2 + C, \ C = 0$

$v(t) = 6t - 2t^2$

$= \begin{cases} s(2) - s(0) = \int_0^2 v(t)dt = \int_0^2 (6t - 2t^2)dt = \left[3t^2 - \dfrac{2}{3}t^3 \right]_0^2 = 12 - \dfrac{16}{3} = \dfrac{20}{3} \\ f(2) - f(0) = \int_0^2 f'(t)dt \end{cases}$

47. (B)

$\begin{cases} x(2) - x(0) = \int_0^2 v(t)dt \\ f(2) - f(0) = \int_0^2 f'(t)dt \end{cases}$

$\Rightarrow \ x(2) = x(0) + \int_0^2 (e^t + t)dt = 3 + \left[e^t + \dfrac{1}{2}t^2 \right]_0^2 = 3 + e^2 + 2 - 1 = 4 + e^2$

48. (B)

Velocity를 $V(t)$, Acceleration을 $A(t)$라고 하면,

$V(t) = 6t^2 - 42t + 72$, $A(t) = 12t - 42$

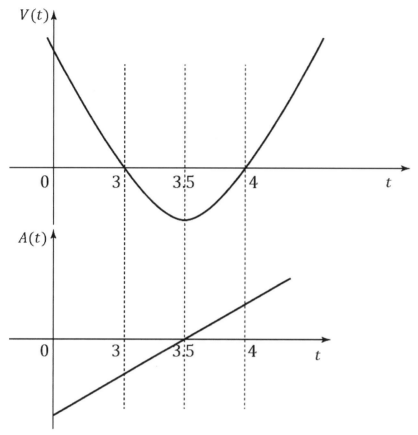

위의 그림에서 보는 바와 같이 $3 < t < 3.5$, $t > 4$에서 velocity와 acceleration의 sign이 같으므로 speed가 increasing 한다.

Topic 12 - Area & Volume

① Area

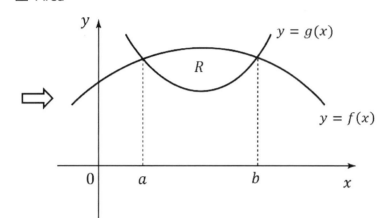

$$R = \int_a^b (f(x) - g(x))dx$$

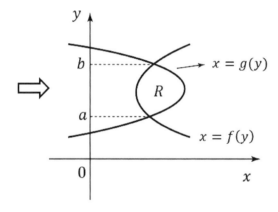

$$R = \int_a^b (g(y) - f(y))dy$$

② Volume

(1) Cross Section Volume

① $\text{Volume} = \int_a^b (Cross\ Section\ Area)dx$

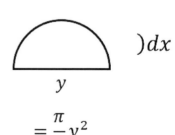

$$= \int_a^b (\quad)dx$$

Square	Equilateral	Semicircle
$= y^2$	$= \dfrac{\sqrt{3}}{4} y^2$	$= \dfrac{\pi}{8} y^2$

② Volume $= \displaystyle\int_a^b (\, Cross\ Section\ Area\,)dy$

$$= \int_a^b (\qquad\qquad\qquad\qquad\qquad\qquad\qquad\qquad)dy$$

Square	Equilateral	Semicircle
□	△	⌓
x	x	x
$= x^2$	$= \dfrac{\sqrt{3}}{4}x^2$	$= \dfrac{\pi}{8}x^2$

(2) Revolution

① Washer

⇒ x축 회전

Volume $= \pi \displaystyle\int_a^b ({y_1}^2 - {y_2}^2)dx$

⇒ y축 회전

Volume $= \pi \displaystyle\int_a^b ({x_1}^2 - {x_2}^2)dy$

② Shell Method

⇒ x축 회전

Volume $= 2\pi \displaystyle\int_a^b xy\,dy$

⇒ y축 회전

Volume $= 2\pi \displaystyle\int_a^b xy\,dx$

Exercise

49. The base of a solid is the region enclosed by a triangle whose vertices are $(0, 0)$, $(2, 0)$, and $(0, 1)$. For this solid, each cross section perpendicular to the y-axis is a rectangle whose height is 2 times the length of its base. What is the volume of the solid?

(A) $\dfrac{4}{3}$　　　　　(B) $\dfrac{8}{3}$　　　　　(C) 3　　　　　(D) $\dfrac{10}{3}$

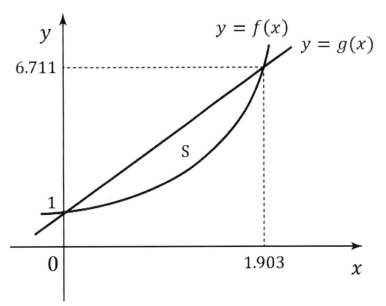

50. Let f and g be the functions given by $f(x) = e^x$ and $g(x) = 3x + 1$. Let S be the region in the first quadrant enclosed by the graphs of f and g as shown in the figure above. What is the volume of the solid generated when S is revolved about the horizontal line $y = -1$?

(A) 14.715　　　　(B) 22.342　　　　(C) 34.138　　　　(D) 46.229

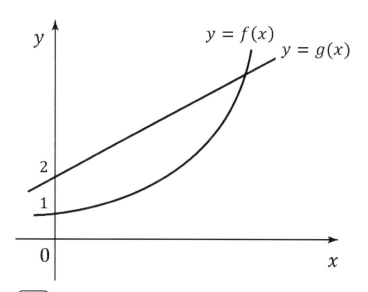

51. Let f and g the functions given by $f(x) = e^x$ and $g(x) = x + 2$. Let S be the region in the first quadrant enclosed by the graphs of f and g as shown in the figure above. The region S is the base of a solid. For this solid, the cross sections perpendicular to the x-axis are equilaterals with diameters extending from $y = f(x)$ to $y = g(x)$. What is the volume of this solid?

(A) 0.285   (B) 0.524   (C) 0.659   (D) 0.887

☝️ Solution

49. (B)

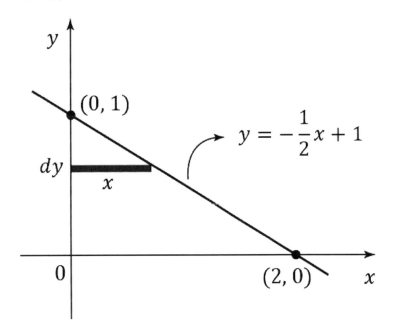

$$\text{Volume} = \int_0^1 \left(\boxed{} \; 2x\right)dy = \int_0^1 2x^2 dy$$

$$\underset{x}{}$$

$\Rightarrow \; y = -\dfrac{1}{2}x + 1$ 에서 $x = 2 - 2y$ 이므로

$\text{Volume} = 2\displaystyle\int_0^1 (2-2y)^2 dy, \; 2-2y = u, \; -2 = \dfrac{du}{dy}$

$\Rightarrow 2\displaystyle\int_2^0 u^2 (-\dfrac{1}{2})du = \int_0^2 u^2 du = \left[\dfrac{1}{3}u^3\right]_0^2 = \dfrac{8}{3}$

50. (D)

회전축을 y축으로 $+1$만큼 이동시켜 회전축이 $x-$axis가 되도록 한다.

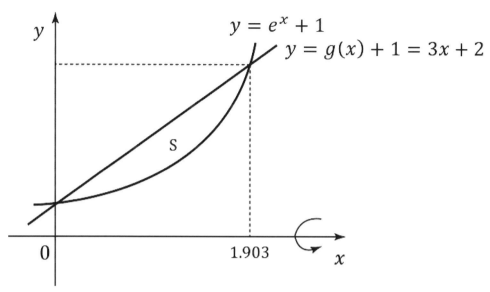

Volume $= \pi \int_0^{1.903} \{(3x+2)^2 - (e^x+1)^2\}dx \approx 46.229$

51. (A)

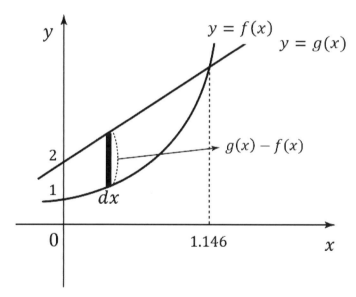

$\text{Volume} = \int_0^{1.146} (\triangle_y)dx = \int_0^{1.146} \frac{\sqrt{3}}{4}y^2 dx$

$= \frac{\sqrt{3}}{4}\int_0^{1.146}(x+2-e^x)^2 dx \approx 0.285$

Topic 13 - Differential Equation

$\Rightarrow \dfrac{dy}{dx} = kx$ 와 같이 $\dfrac{dy}{dx}$ 를 보고 y 를 구하는 방법은

① 같은 Variable끼리 모은다.

② 양변에 $\displaystyle\int$ 을 취한다.

⚲ Exercise

52. If $f(0)=1$, $\dfrac{dy}{dx}=\dfrac{xe^{x^2}}{y}$ and $y>0$ for all x, what is y?

(A) $y=\sqrt{e^{x^2}}$

(B) $y=-\sqrt{e^{x^2}}$

(C) $y=\pm\sqrt{e^{x^2}}$

(D) $y=e^{x^2}$

Solution

52. (A)

① 우선 같은 Variable끼리 모은다.

$\Rightarrow ydy = xe^{x^2}dx$

② 양변에 $\int$ 를 취한다.

$\Rightarrow \int ydy = \int xe^u \cdot \dfrac{1}{2x}du,\ x^2 = u,\ 2x = \dfrac{du}{dx}$

$\Rightarrow \dfrac{1}{2}y^2 = \int xe^u \cdot \dfrac{1}{2x}du \Rightarrow \dfrac{1}{2}y^2 = \dfrac{1}{2}e^u + C$

$\Rightarrow \dfrac{1}{2}y^2 = \dfrac{1}{2}e^{x^2} + C$ 에서 $(0,\ 1)$을 대입하면 $C = 0$.

그러므로 $y^2 = e^{x^2}$에서 $y = \pm\sqrt{e^{x^2}}$.

③ $(0, 1)$ 조건으로부터 $+$인지 $-$인지를 결정한다.

$1 = \pm\sqrt{e^0}$ 에서 $1 = +\sqrt{e^0} = \sqrt{1}$

그러므로, $y = \sqrt{e^{x^2}}$

Topic 14 - ①②③④

① The Extreme Value Theorem
② The Intermediate Value Theorem (IVT)
③ Mean Value Theorem (MVT)
④ Rolle's Theorem

① The Extreme Value Theorem

$y = f(x)$가 주어진 구간에서 연속이면 주어진 구간 내에서 maximum, minimum이 존재한다.

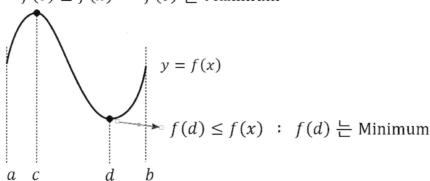

$$f(c) \geq f(x) \ : \ f(c) \text{ 는 Maximum}$$

$y = f(x)$

$$f(d) \leq f(x) \ : \ f(d) \text{ 는 Minimum}$$

$a \quad c \qquad d \quad b$

② The Intermediate Value Theorem

$y = f(x)$가 주어진 구간에서 연속이면

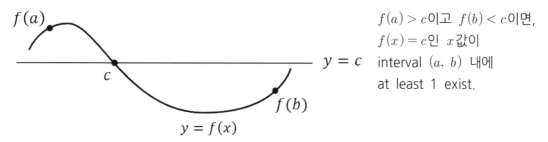

$f(a)$

c

$y = c$

$f(b)$

$y = f(x)$

$f(a) > c$이고 $f(b) < c$이면,
$f(x) = c$인 x값이
interval (a, b) 내에
at least 1 exist.

⇒ The Extreme Value Theorem의 경우 $y = f(x)$가 주어진 구간에서 continuous라는 조건만 있으면 되지만 Intermediate Value Theorem은 $f(a)$와 $f(b)$값까지도 알아야 한다.

③ Mean Value Theorem

$y = f(x)$가 주어진 구간에서 "Differentiable"이면 다음의 식이 성립한다.

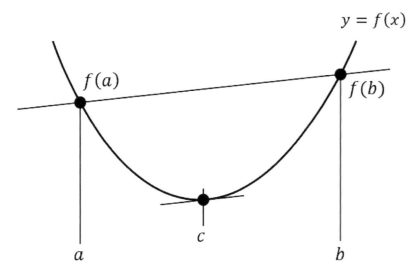

$\dfrac{f(b) - f(a)}{b - a} = f'(c)$를 만족하는 c가 구간 $a < x < b$ 내에 at least 1 exist 한다.

④ Rolle's Theorem

$y = f(x)$가 주어진 구간에서 "Differentiable"이면서 $f(a) = f(b)$이면 다음의 식이 성립한다.

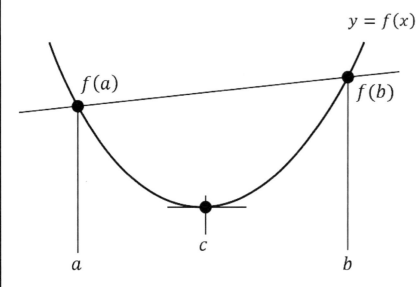

$\dfrac{f(b) - f(a)}{b - a} = 0 = f'(c)$를 만족하는 c가 구간 $a < x < b$ 내에 at least 1 exist 한다.

53. If f is a continuous function on the closed interval $[1, 5]$, which of the following must be true?

(A) There is a number c in the open interval $(1, 5)$ such that $f'(c) = \dfrac{f(5) - f(1)}{4}$.

(B) There is a number c in the open interval $(1, 5)$ such that $f'(c) = 0$.

(C) There is a number c in the open interval $(1, 5)$ such that $f(c) = 0$.

(D) There is a number c in the open interval $(1, 5)$ such that $f(c) \leq f(x)$ for all x in $[1, 5]$.

x	1	2	3	4	5	6
$f(x)$	5	1	-1	0	-3	5

54. The function f is continuous and differentiable on the closed interval $[1, 6]$, The table above gives selected values of f on this interval. Which of the following statements must be true?

(A) f have three inflection points.

(B) The maximum value of f on $[1, 6]$ is 5.

(C) There is a number c in the open interval $(1, 6)$ such that $f'(c) = 0$.

(D) $f(x) > 0$ for $1 < x < 2$.

Solution

53. (D)

(A) $f'(c) = \dfrac{f(5) - f(1)}{5 - 1}$ ⇒ 문제에서 "Differentiable" 조건이 없으므로 Mean Value Theorem이 성립하지 않는다.

(B) 문제에서 "Differentiable"과 "$f(1) = f(5)$" 조건이 없으므로 Rolle's Theorem이 성립하지 않는다.

(C) 문제에서 $f(1)$, $f(5)$ 값을 알 수 없으므로 Intermediate Value Theorem이 성립하지 않는다.

(D) Continuous 조건이 있으므로 주어진 구간 내에서 function f는 minimum value ($f(c)$)를 갖는다.

54. (C)

문제에서 주어진 Table로 Graph를 그려보면 여러 가지 상황이 나올 수 있다. 예를 들어, 한 가지 경우를 그려보면 다음과 같을 수 있다.

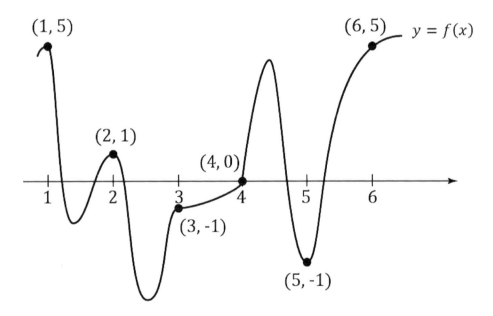

(A) Inflection Point가 몇 개일지는 알 수 없다.

(B) Maximum Value가 5보다 더 클 수도 있다.

(C) Function f는 differentiable이고 $f(1) = f(6)$이므로 Rolle's Theorem이 성립한다.

(D) $1 < x < 2$에서 반드시 $f(x) > 0$이라고 할 수 없다.

Topic 15 - Line Tangent Equation

I. Tangents

접선의 방정식(The equation of the tangent line)은 다음과 같이 구한다.

① Slope 구하기 $f'(x_1)$, $\dfrac{dy}{dx}$

⇒ ② 지나는 점 (Tangency) 대입.

즉, $\boxed{y - y_1 = f'(x_1)(x - x_1)}$

II. Normals

Normal Line 방정식은 다음과 같이 구한다.

① Tangent Line Slope 구하기

$f'(x_1)$, $\dfrac{dy}{dx}$

② Normal Line Slope 구하기

$-\dfrac{1}{f'(x_1)}$, $-\dfrac{dx}{dy}$

(※두 직선이 수직(Perpendicular)이면 기울기끼리 곱은 -1이다.

Exercise

55. Let f be a differentiable function with $f(-2)=5$ and $f'(-2)=3$, and let g be the function defined $g(x)=x^3 \cdot f(x)$. What is the equation of the line tangent to the graph of g at the point where $x=-2$?

(A) $y=36(x-2)+40$

(B) $y=36(x+2)-40$

(C) $y=8(x+2)+20$

(D) $y=-8(x+2)+20$

56. At what point on the graph of $y=x^2$ is the tangent line parallel to the line $2x-y=5$?

(A) $(0, 0)$ (B) $(0, -1)$ (C) $(1, 1)$ (D) $(-1, 2)$

Solution

55. (B)

① Slope : $g'(x) = 3x^2 f(x) + x^3 f'(x)$

$\Rightarrow g'(-2) = 3 \cdot (-2)^2 \cdot f(-2) - 8f'(-2) = 60 - 8 \times 3 = 36$

② $(-2, g(-2))$를 지나므로 $g(-2) = -8f(-2) = -8 \times 5 = -40$

그러므로, $(-2, -40)$를 지난다.

$y + 40 = 36(x+2) \Rightarrow y = 36(x+2) - 40$

56. (C)

$y = 2x + 5$이므로 Slope는 2.

$y' = 2x = 2$에서 $x = 1$, $y = 1$

그러므로, $(1, 1)$

$$\textbf{Topic 16 - } \lim_{n\to\infty}\sum_{k=1}^{n}f(k) \Rightarrow \int_{a}^{b}f(x)dx$$

$\displaystyle\lim_{n\to\infty}\sum_{k=1}^{n}f(k)$를 $\displaystyle\int_{a}^{b}f(x)dx$로 바꿀 때는 다음과 같이 한다.

① Graph를 대략 그린다. 일단 $y=f(x)$의 Graph 모양을 모르기 때문에 대략 그리는 것이다.

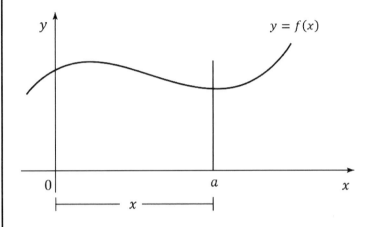

② 범위는 내 마음대로 잡아도 된다. Origin으로부터의 거리가 x값이므로 항상 Origin으로부터 범위를 잡도록 한다.

③

길이 a를 n등분 $= \dfrac{a}{n} = dx$

④ Shaded 된 부분의 Area는 ▌를 무수히 많이 더한 것이다.

⇒ · $\lim\limits_{n \to \infty} \sum\limits_{k=1}^{n} = \int$

· $\dfrac{a}{n} = dx$

· $\dfrac{a}{n} \cdot k = x$

다음의 예제를 이와 같은 순서대로 3가지 방법으로 구해보자.

(Example) Evaluate $\lim\limits_{n \to \infty} \sum\limits_{k=1}^{n} (1 + \dfrac{2k}{n}) \dfrac{1}{n}$

(Solution ①)

① Graph를 대략 그리고 범위를 Origin으로부터 1까지 잡는다.

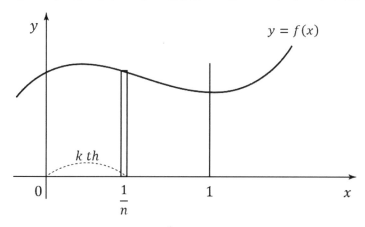

② · 길이 1을 n등분 = $\dfrac{1}{n}$

· $\dfrac{1}{n} \cdot k = x$

· $\lim\limits_{n \to \infty} \sum\limits_{k=1}^{n} = \int$

③

$$\boxed{\lim\limits_{n \to \infty} \sum\limits_{k=1}^{n}} \left(1 + 2 \cdot \boxed{\dfrac{k}{n}}\right) \cdot \boxed{\dfrac{1}{n}}$$
$$= \int_{0}^{1} \qquad = x \quad = dx$$

$\Rightarrow \displaystyle\int_{0}^{1} (1 + 2x)dx, \ 1 + 2x = u$

$\Rightarrow \dfrac{1}{2} \displaystyle\int_{1}^{3} u\,du = \dfrac{1}{2}\left[\dfrac{1}{2}u^2\right]_{1}^{3} = \dfrac{1}{2}(\dfrac{9}{2} - \dfrac{1}{2}) = \dfrac{8}{4} = 2$

(Solution ②)
① Graph를 대략 그리고 범위를 Origin으로부터 2까지 잡는다.

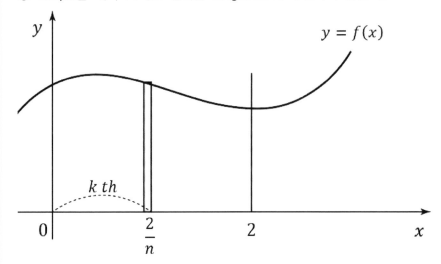

② · 길이 2를 n등분 $\Rightarrow \dfrac{2}{n} = dx \Rightarrow \dfrac{1}{n} = \dfrac{1}{2}dx$

· $\dfrac{2}{n} CDOT k = x$

· $\displaystyle\lim_{n \to \infty} \sum_{k=1}^{n} = \int$

③

$$\boxed{\lim_{n \to \infty} \sum_{k=1}^{n}} \left(1 + \boxed{2 \cdot \dfrac{k}{n}} \right) \cdot \boxed{\dfrac{1}{n}}$$
$$= \int_0^2 \qquad = x \qquad = \dfrac{1}{2}dx$$

$\Rightarrow \dfrac{1}{2}\displaystyle\int_0^2 (1+x)dx = \dfrac{1}{2}\left[x + \dfrac{1}{2}x^2 \right]_0^2 = \dfrac{1}{2}\left(2 + \dfrac{4}{2} \right) = \dfrac{1}{2} \times 4 = 2$

(Solution ③)
① Graph를 대략 그리고 범위를 Origin으로부터 3까지 잡는다.

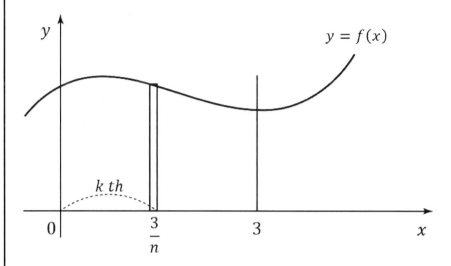

② • 길이 3을 n등분 $\Rightarrow \dfrac{3}{n} = dx \Rightarrow \dfrac{1}{n} = \dfrac{1}{3}dx$

 • $\dfrac{3}{n} CDOT k = x \Rightarrow \dfrac{2}{n} \cdot k = \dfrac{2}{3}x$

 • $\displaystyle\lim_{n\to\infty}\sum_{k=1}^{n} = \int$

③

$$\boxed{\lim_{n\to\infty}\sum_{k=1}^{n}}\left(1+\boxed{\dfrac{2}{n}k}\right)\cdot\boxed{\dfrac{1}{n}}$$

$$= \int_{0}^{3} \qquad = \dfrac{2}{3}x \qquad = \dfrac{1}{3}dx$$

$\Rightarrow \dfrac{1}{3}\displaystyle\int_{0}^{3}\left(1+\dfrac{2}{3}x\right)dx,\ 1+\dfrac{2}{3}x = u \Rightarrow \dfrac{2}{3} = \dfrac{du}{dx}$

$\Rightarrow \dfrac{1}{3}\displaystyle\int_{1}^{3} u \cdot \dfrac{3}{2}du = \dfrac{1}{2}\left[\dfrac{1}{2}u^2\right]_{1}^{3} = \dfrac{1}{2}\left(\dfrac{9}{2}-\dfrac{1}{2}\right) = \dfrac{1}{2}\cdot 4 = 2$

※Solution ①, ②, ③과 같이 범위를 어디까지 설정하더라도 결과는 똑같이 나온다.

57. $\displaystyle\lim_{n\to\infty}\sum_{k=1}^{n}\cos\left(\frac{\pi}{6}+\frac{\pi k}{4n}\right)\cdot\frac{1}{n}=$

(A) $\displaystyle\int_{0}^{\frac{\pi}{4}}\cos\left(\frac{\pi}{6}+x\right)dx$

(B) $\displaystyle\frac{4}{\pi}\int_{0}^{\frac{\pi}{4}}\cos\left(\frac{\pi}{6}+x\right)dx$

(C) $\displaystyle\frac{\pi}{4}\int_{0}^{\frac{\pi}{4}}\cos\left(\frac{\pi}{6}+\frac{\pi}{4}x\right)dx$

(D) $\displaystyle\frac{4}{\pi}\int_{0}^{\frac{\pi}{4}}\cos x\,dx$

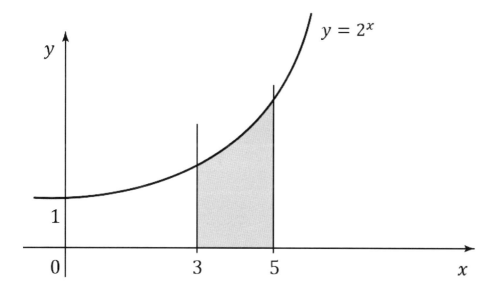

58. The function f is given by $f(x) = 2^x$. The graph f is shown above. Which of the following limits is equal to the area of shaded region?

(A) $\displaystyle\lim_{n \to \infty} \sum_{k=1}^{n} (2^{3 + \frac{2}{n}k}) \cdot \frac{2}{n}$

(B) $\displaystyle\lim_{n \to \infty} \sum_{k=1}^{n} (2^{\frac{2}{n}k}) \cdot \frac{2}{n}$

(C) $\displaystyle\lim_{n \to \infty} \sum_{k=1}^{n} (2^{3 + \frac{2}{n}k}) \cdot \frac{1}{n}$

(D) $\displaystyle\lim_{n \to \infty} \sum_{k=1}^{n} (2^{\frac{2}{n}k}) \cdot \frac{1}{n}$

59. Which of the following limits is equal to $\displaystyle\int_1^4 e^x dx$?

(A) $\displaystyle\lim_{n\to\infty}\sum_{k=1}^{n} e^{\frac{3}{n}k} \cdot \frac{1}{n}$

(B) $\displaystyle\lim_{n\to\infty}\sum_{k=1}^{n} (e^{1+\frac{3}{n}k}) \cdot \frac{1}{n}$

(C) $\displaystyle\lim_{n\to\infty}\sum_{k=1}^{n} e^{\frac{3}{n}k} \cdot \frac{3}{n}$

(D) $\displaystyle\lim_{n\to\infty}\sum_{k=1}^{n} (e^{1+\frac{3}{n}k}) \cdot \frac{3}{n}$

Solution

57. (B)

① $y = f(x)$를 그린다. 범위는 Origin부터 $\frac{\pi}{4}$까지 잡는다. (※범위는 앞에서 설명하였듯이 임의로 잡아도 된다.)

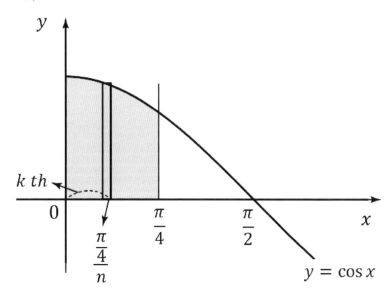

$$\Rightarrow \bullet \ \frac{\frac{\pi}{4}}{n} = \frac{\pi}{4n} = dx \ \Rightarrow \ \frac{1}{n} = \frac{4}{\pi} dx$$

$$\bullet \ \frac{\pi}{4n} \times k = x$$

$$\bullet \ \lim_{n \to \infty} \sum_{k=1}^{n} = \int_{0}^{\frac{\pi}{4}}$$

그러므로, $\boxed{\lim_{n \to \infty} \sum_{k=1}^{n}} \cos\left(\frac{\pi}{6} + \boxed{\frac{\pi k}{4n}}\right) \cdot \boxed{\frac{1}{n}}$

$$= \int_{0}^{\frac{\pi}{4}} \qquad \underset{= x}{} \qquad = \frac{4}{\pi} dx$$

$$\Rightarrow \frac{4}{\pi} \int_{0}^{\frac{\pi}{4}} \cos\left(\frac{\pi}{6} + x\right) dx$$

58. (A)

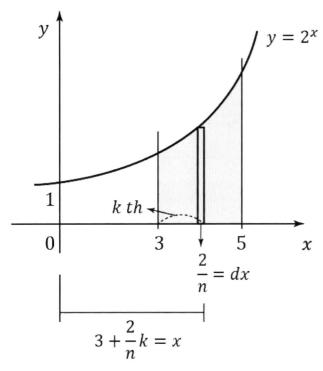

$$\Rightarrow \lim_{n \to \infty} \sum_{k=1}^{n} (2^{3 + \frac{2}{n}k}) \cdot \frac{2}{n}$$

59. (D)

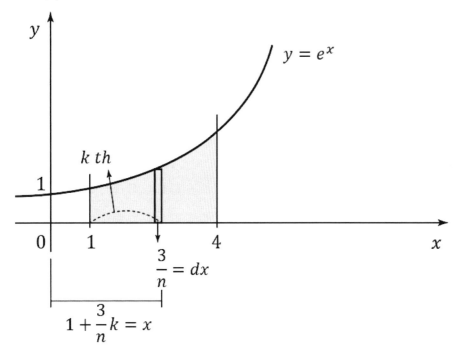

$$\Rightarrow \lim_{n \to \infty} \sum_{k=1}^{n} (e^{1 + \frac{3}{n}k}) \cdot \frac{3}{n}$$

Topic 17 - Riemann Sum

$y = f(x)$의 곡선과 x축 사이의 면적을 사각형으로 분할하고 더하는 방법으로 구하는 것을 "Riemann Sum", 사다리꼴로 분할하고 더하는 방법으로 구하는 것을 "Trapezoid Rule"이라고 한다.
이 단원에서 암기할 것은 없다.

Riemann Sum은 사각형의 높이가 Left-endpoint인지 Right-endpoint인지 Midpoint인지에 따라 3가지로 나뉜다. 다음의 예제를 통해 Riemann Sum과 Trapezoid Rule을 이해하도록 하자.

(Example) Find the approximate area under the curve of $f(x) = x^2 + 1$ from $x = 0$ to $x = 9$, using 3 left-endpoint rectangles.

(Solution)

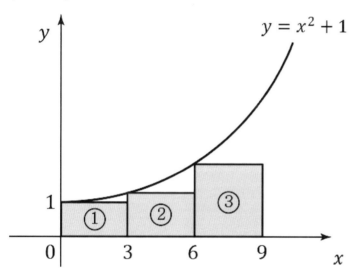

Left
$\downarrow$
$$L(3) = \int_0^9 (x^2 + 1)dx = \underbrace{3 \times f(0)}_{① \text{면적}} + \underbrace{3 \times f(3)}_{② \text{면적}} + \underbrace{3 \times f(6)}_{③ \text{면적}} = 3\big(f(0) + f(3) + f(6)\big)$$
$\uparrow$
Sub-interval

$$= 3(1 + 10 + 37) = 144$$

(Example) Find the approximate area under the curve of $f(x) = x^2 + 1$ from $x = 0$ to $x = 9$, using 3 right-endpoint rectangles.

(Solution)

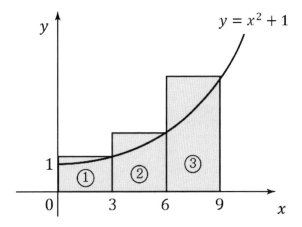

Right
$$
R(3) = \int_0^9 (x^2 + 1)dx = \underbrace{3 \times f(3)}_{① 면적} + \underbrace{3 \times f(6)}_{② 면적} + \underbrace{3 \times f(9)}_{③ 면적} = 3\big(f(3) + f(6) + f(9)\big)
$$
Sub-interval
$$
= 3(10 + 37 + 82) = 387
$$

(Example) Find the approximate area under the curve of $f(x) = x^2 + 1$ from $x = 0$ to $x = 9$, using 3 midpoint rectangles.

(Solution)

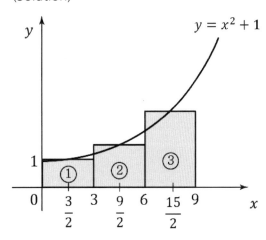

$$
M(3) = \int_0^9 (x^2 + 1)dx = \underbrace{3 \times f\left(\frac{3}{2}\right)}_{① 면적} + \underbrace{3 \times f\left(\frac{9}{2}\right)}_{② 면적} + \underbrace{3 \times f\left(\frac{15}{2}\right)}_{③ 면적}
$$
$$
= 3\left(f\left(\frac{3}{2}\right) + f\left(\frac{9}{2}\right) + f\left(\frac{15}{2}\right)\right) = 3\left(\frac{13}{4} + \frac{85}{4} + \frac{229}{4}\right) = 245.25
$$

(Example) Find the approximate area under the curve of $f(x) = x^2 + 1$ from $x = 0$ to $x = 9$, using 3 trapezoids.

(Solution)

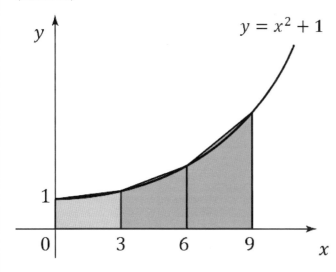

Trapezoids
↓
$$T(3) = \int_0^9 (x^2 + 1)dx = \frac{1}{2} \times 3 \times \left(f(0) + f(3)\right) + \frac{1}{2} \times 3 \times \left(f(3) + f(6)\right) + \frac{1}{2} \times 3 \times \left(f(6) + f(9)\right)$$
↑
Sub-interval
$$= 3(f(0) + 2f(3) + 2f(6) + f(9)) = 265.5$$

Exercise

60. Let f be a function that is continuous for all real numbers. The table below gives values of f for selected point in the closed interval $[1, 12]$.

x	1	3	5	7	9	11
$f(x)$	2	4	2	-4	2	6

Use a trapezoidal sum with sub-intervals indicated by the data in the table to approximate $\int_{1}^{11} f(x)dx$.

(A) 8   (B) 14   (C) 16   (D) 24

Solution

60. (C)
다음과 같이 그려서 해결한다.

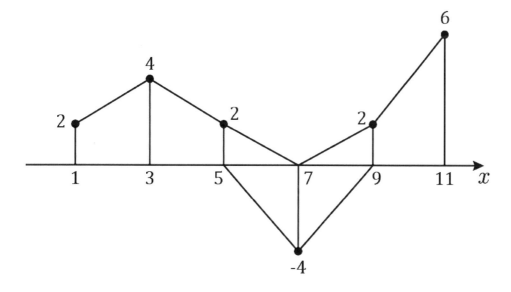

$$T = \frac{1}{2} \times 2 \times (2+4) + \frac{1}{2} \times 2 \times (4+2) + \frac{1}{2} \times 2 \times (2-4) + \frac{1}{2} \times 2 \times (-4+2) + \frac{1}{2} \times 2 \times (2+6)$$

$$= 16$$

Topic 18 - Slope Field

"Direction Field"라고도 한다. 공부하기에 쉬운 단원 중 하나이다.

예를 들어, $\dfrac{dy}{dx} = 2x$ 라고 할 때, 주어진 좌표에 Slope Field를 나타내보자.

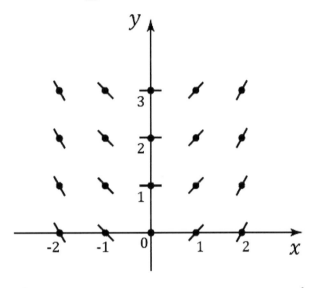

$\dfrac{dy}{dx}$ 는 Slope를 나타낸다. 좌표가 $(2, 3)$ 이므로 $\dfrac{dy}{dx} = 2 \times 2 = 4$. 즉, $(2, 3)$ 에서의 접선(The tangent line)

앞의 그림에서 수많은 선분(Segment)들은 각각의 점에서의 접선(The tangent line)이며 이 수많은 접선들을 이용해 굳이 Differential Equation을 풀지 않고도 원래 함수의 그래프(Solution Curve)의 형태를 짐작할 수 있다. 그렇다면, 위의 Slope Field로부터 $(0, 1)$ 을 지나는 Solution Curve를 그려보자.

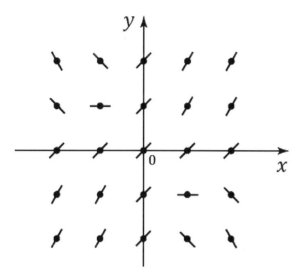

Exercise

60. Shown above is a slope field for which of the following differential equation?

(A) $\dfrac{dy}{dx} = xy + 1$

(B) $\dfrac{dy}{dx} = x^2 + y + 1$

(C) $\dfrac{dy}{dx} = xy^2$

(D) $\dfrac{dy}{dx} = x + y + 1$

⚡ Solution

61. (A)

각 점마다의 간격을 1이라 가정하고 주어진 Choice에 대입하여 본다.

AP Calculus AB
☀ Practice Test 1

CALCULUS AB
SECTION I, Part A
Time-60 minutes
Number of questions-30

A CALCULATOR MAY NOT BE USED ON THIS PART OF THE EXAMINATION.

Directions: Solve each of the following problems, using the available space for scratchwork. After examining the form of the choices, decide which is the best of the choices given and fill in the corresponding oval on the answer sheet. No credit will be given for anything written in the test book. Do not spend too much time on any one problem.

In this test:

(1) Unless otherwise specified, the domain of a function is assumed to be the set of all real numbers x for which $f(x)$ is a real number.

(B) The inverse of a trigonometric function f may be indicated using the inverse function notation f^{-1} or with the prefix "arc" (e.g., $\sin^{-1} = \arcsin x$).

1. $\int x^3 \cos(x^4) dx =$

(A) $\frac{1}{4}\cos x^4 + C$

(B) $\frac{1}{4}\cos x + C$

(C) $\frac{1}{4}\sin(x^4) + C$

(D) $\frac{1}{4}\sin x + C$

2. $\lim\limits_{x \to 1} \dfrac{\displaystyle\int_1^x e^{t^5} dt}{3x^3 - 3}$ is

(A) $\dfrac{e}{9}$

(B) 9

(C) e

(D) $9e$

3. If $(xy + y^2)^3 = 8x$, then $x = y = 1$, what is $\dfrac{dy}{dx}$?

(A) $-\dfrac{1}{9}$

(B) $-\dfrac{1}{3}$

(C) 1

(D) 3

4. If $f(x) = 3x^2 + 1$ and $g(x) = \cos x$, then $(g \circ f)'$ is

(A) $-6x \cos(3x^2 + 1)$

(B) $6 \sin(3x^2 + 1)$

(C) $\sin(3x^2 + 1)$

(D) $-6x \sin(3x^2 + 1)$

5. If f is a continuous function and if $F'(x) = f(x)$ for all real numbers x,

then $\displaystyle\int_{1}^{2} f(5x)\,dx =$

(A) $F(10) - F(5)$

(B) $\dfrac{1}{5}(F(10) - F(5))$

(C) $\dfrac{1}{5}(F(2) - F(1))$

(D) $F(2) - F(1)$

6. If $f''(x) = (x-1)^2(x+1)(x+3)^3$, then the graph of $f(x)$ has inflection point(s) when $x =$
(A) -3
(B) -1
(C) -1 and 1
(D) -3 and -1

7. The graph of the function $y = x^3 + 6x^2$ changes concavity at $x =$

(A) -2

(B) -1

(C) 0

(D) 1

8. The base of a solid is the region in the first quadrant by the parabola $y = e^x$, the line $x = 3$, and the x-axis. Each plane section of the solid perpendicular to the x-axis is square. What is the volume of the solid?

(A) $2(e^3 - 1)$

(B) $\dfrac{1}{2}(e^3 - 1)$

(C) $e^6 - 1$

(D) $\dfrac{1}{2}(e^6 - 1)$

x	1	2	3	4	5
$f''(x)$	3	4	1	-1	-2

9. The polynomial function f has selected values of its second derivative f'' given in the table above. Which of the following statements must be true?

(A) f has relative maximum at $x = 2$.

(B) f is concave down on the interval $(1, 3)$.

(C) f has at least one inflection point on the interval $(3, 4)$.

(D) f is decreasing on the interval $(4, 5)$.

x	$f(x)$	$f'(x)$
1	1	2
2	3	5
3	4	6

10. The table above gives selected values for a differentiable function f and its derivative. If g is the inverse function of f, what is the value of $g'(3)$?

(A) $\dfrac{1}{5}$

(B) $\dfrac{1}{6}$

(C) $\dfrac{1}{2}$

(D) 3

11. What is the solution of the differential equation $\dfrac{dy}{dx} = x \cdot \cos x^2$ when $y(0) = 1$?

(A) $y = \dfrac{1}{2}\cos(2x) + 1$

(B) $y = \dfrac{1}{2}\sin(2x) + 1$

(C) $y = \dfrac{1}{2}\cos x^2 + 1$

(D) $y = \dfrac{1}{2}\sin x^2 + 1$

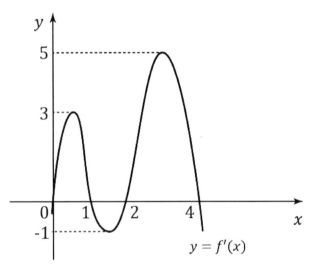

$y = f'(x)$

12. The graph of f', the derivative of the function f, is shown above. Which of the following must be true?

I. $f(4) > f(1)$
II. $f(2) > f(0)$
III. $f(2) > f(1)$

(A) I only
(B) II only
(C) I and II only
(D) II and III only

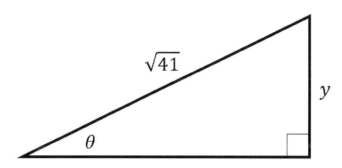

13. In the triangle shown above, if θ increases at a constant rate of 5 radians per second, at what rate is y increasing in units per minute when y equals 4 units?

(A) 5

(B) $\dfrac{5}{\sqrt{41}}$

(C) 25

(D) $\sqrt{41}$

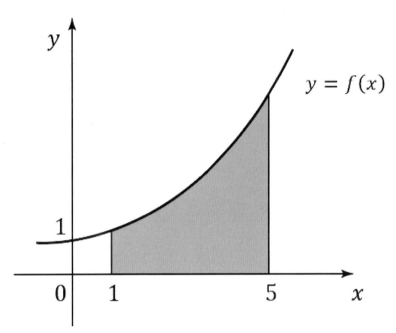

14. The function f is given by $f(x) = 3^x$. The graph f is shown above. Which of the following limits is equal to the area of the shaded region?

(A) $\lim\limits_{n \to \infty} \sum\limits_{k=1}^{n} 3^{\frac{4}{n}k} \cdot \dfrac{5}{n}$

(B) $\lim\limits_{n \to \infty} \sum\limits_{k=1}^{n} 3^{1+\frac{4}{n}k} \cdot \dfrac{5}{n}$

(C) $\lim\limits_{n \to \infty} \sum\limits_{k=1}^{n} 3^{1+\frac{4}{n}k} \cdot \dfrac{4}{n}$

(D) $\lim\limits_{n \to \infty} \sum\limits_{k=1}^{n} 3^{\frac{4}{n}k} \cdot \dfrac{4}{n}$

15. Let f be a differentiable function with $f(1)=3$ and $f'(1)=2$, and let g be the function defined $g(x)=x^3 \cdot f(x)$. What is the equation of the line tangent to the graph of g at the point where $x=1$?

(A) $g(x)=11x$

(B) $g(x)=11x-8$

(C) $g(x)=9x-6$

(D) $g(x)=9x$

x	$f(x)$	$f'(x)$	$f''(x)$	$f'''(x)$
5	0	0	0	2

16. The third derivative of the function f is continuous on the interval $(0, 6)$. Values for f and its first three derivatives at $x=5$ are given in the table above. What is $\lim\limits_{x\to 5}\dfrac{2f(x)}{(x-5)^3}$?

(A) 0

(B) $\dfrac{2}{3}$

(C) 1

(D) The limit does not exist

17. The function f is continuous for $1 \le x \le 5$ and differentiable for $1 < x < 5$. If $f(1) = -3$ and $f(5) = 3$, which of the following statements could be false?

(A) There exists c, where $1 < c < 5$, such that $f(c) = 0$.

(B) There exists c, where $1 \le c \le 5$, such that $f(c) \le f(x)$ for all x on the closed interval $1 \le x \le 5$.

(C) There exists c, where $1 < c < 5$, such that $f'(c) = \dfrac{3}{2}$.

(D) There exists c, where $1 < c < 5$, such that $f'(c) = 0$.

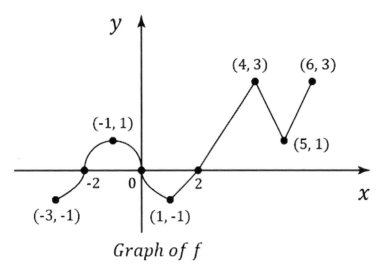

(4, 3)   (6, 3)

(-1, 1)

(5, 1)

-2   0   2

(-3, -1)   (1, -1)

Graph of f

18. The function f is defined on the closed interval $[-3, 6]$. Let g be the function defined by $g(x) = \displaystyle\int_{-3}^{x} f(t)\,dt$. Which of the following statements is true about g?

(A) g is not differentiable at $x = 1, 4,$ and 5.

(B) g has a local minimum at $x = 0$.

(C) g is concave down and decreasing for $0 < x < 1$.

(D) g is increasing and concave up for $1 \le x \le 6$.

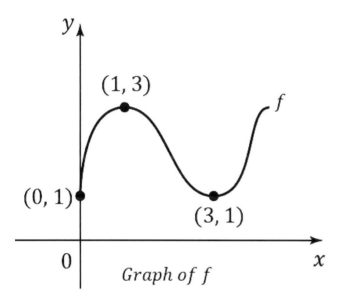

Graph of f

19. The graph of the function f is shown in the figure above. If $h(x) = \int_{1}^{3x} f(t)dt$, what is the value of $h'(1)$?

(A) 3
(B) 6
(C) 9
(D) 12

Time t (minute)	1	2	3	4	5
Temperature $H(t)$ (°C)	17	22	36	45	78

20. The table above shows the change in temperature over time as water is boiled in a pot. The function $H(t)$ is increasing and twice differentiable. Which of the following must be true?

(A) $H''(t) > 0$

(B) $\displaystyle\int_1^5 H'(t)dt = 55$

(C) The average temperature of the water using a left Riemann Sum with four subintervals indicated by the data in the table is 30.

(D) $H(t)$ has an inflection point at $t = 2$.

21. $\lim\limits_{x \to 0} \dfrac{3}{1+5^{\frac{1}{x}}} =$

(A) Nonexistent

(B) 0

(C) $\dfrac{1}{2}$

(D) 3

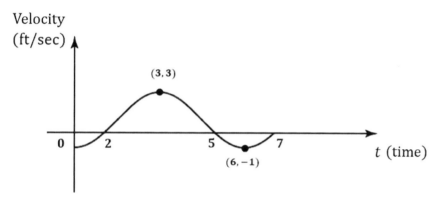

22. The graph above shows the velocity of an object moving along a straight line during the time interval $0 \le t \le 7$. Which of the following must be true?

(A) This object changes its direction at time $t = 6$.

(B) This object has maximum speed at time $t = 2$.

(C) Position graph of this object has inflection point at time $t = 3$, $t = 6$.

(D) Speed of this object is increasing over the time interval $0 \le t \le 2$.

23. If $\tan(x^2 y) = x$, then $\dfrac{dy}{dx}$ is

(A) $\dfrac{\cos^2(x^2 y) - 2xy}{x^2}$

(B) $\dfrac{\cos^2(x^2 y) - 2}{x^2}$

(C) $\dfrac{\sec^2(x^2 y) - x}{x^2}$

(D) $\dfrac{\sec^2(x^2 y) - 2xy}{x^2}$

24. $\displaystyle\int \dfrac{1}{x^2 + 10x + 26}\,dx =$

(A) 1

(B) $\tan^{-1} x + C$

(C) $\sin^{-1}(x+1) + C$

(D) $\tan^{-1}(x+5) + C$

25. If $x^2 + y^2 = 2$, what is the value of $\dfrac{d^2y}{dx^2}$ at the point $(1, 1)$?

(A) -2

(B) -1

(C) $-\dfrac{1}{2}$

(D) 0

26. A water tank is in the shape of an inverted cone. Water is leaking out from the bottom of the inverted cone so that water is falling at the rate of $\dfrac{1}{10}$ ft/sec. If the tank has a height of 6 feet and a radius of 2 feet, at what rate is the water leaking when the water is $\dfrac{2}{3}$ ft in depth?

(A) $\dfrac{\pi}{135}$ ft^3/sec

(B) $\dfrac{2\pi}{135}$ ft^3/sec

(C) $\dfrac{2\pi}{205}$ ft^3/sec

(D) $\dfrac{2\pi}{405}$ ft^3/sec

27. Let f be a function such that $\lim\limits_{h \to 0} \dfrac{f(1+h)-f(1)}{h} = 3$. Which of the following must be true?

I. $\lim\limits_{x \to 1-} f(x) = \lim\limits_{x \to 1+} f(x)$

II. $f(1) = 3$

III. $\lim\limits_{x \to 1} f(x) = 3$

(A) I only

(B) II only

(C) I and III only

(D) II and III only

28. When the region enclosed by the graphs of $y = -x^2 + 2x$ and x-axis is revolved about the y-axis, the volume of the solid generated is given by

(A) $2\pi \displaystyle\int_0^2 (-x^3 + 2x^2)dx$

(B) $\pi \displaystyle\int_0^2 (-x^3 + 2x^2)dx$

(C) $2\pi \displaystyle\int_0^2 (-x^2 + 2x)dx$

(D) $\pi \displaystyle\int_0^2 (-x^2 + 2x)dx$

29. Which of the following must be true?

(A) If function f satisfies $\lim_{x \to 1-} f(x) = \lim_{x \to 1+} f(x)$, it satisfies $\lim_{x \to 1} f(x) = f(1)$.

(B) If function f is continuous at $x = a$, it is unconditionally differentiable at $x = a$.

(C) If function f satisfies $\lim_{x \to 0} \dfrac{f(x) - f(0)}{x} = 3$, it satisfies $\lim_{x \to 0-} f(x) = \lim_{x \to 0+} f(x)$.

(D) For function f, if $f(1) = 3$, it is $\lim_{x \to 1} f(x) = 3$.

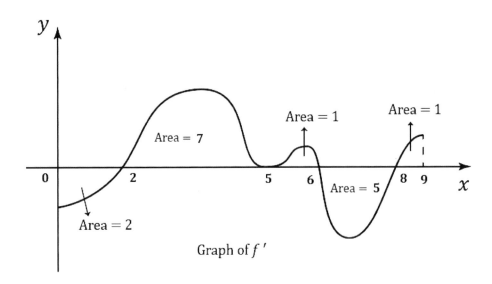

Graph of f'

30. The graph above shows the graph of f', the derivative of a twice-differentiable function f, on the closed interval $0 \le x \le 9$. The area of the regions between the graph of f' and the $x-$ axis are labeled in the figure. The function f is defined for all real numbers and satisfies $f(9) = 5$. What is the absolute minimum value of f on the closed interval $0 \le x \le 9$?

(A) -1

(B) 1

(C) 2

(D) 4

CALCULUS AB
SECTION I, PART B
Time-45 minutes
Number of questions-15

A GRAPHING CALCULATOR IS REQUIRED FOR SOME QUESTIONS ON THIS PART OF THE EXAMINATION.

Directions: Solve each of the following problems, using the available space for scratch work. After examining the form of the choices, decide which is the best of choices given and fill in the corresponding oval on the answer sheet. No credit will be given for anything written in the test book. Do no spend too much time on any one problem.

BE SURE YOU ARE USING PAGE 3 OF THE ANSWER SHEET TO RECORD YOUR ANSWERS TO QUESTIONS NUMBERED 76-90.

YOU MAY NOT RETURN TO PAGE 2 OF THE ANSWER SHEET.

In this test:

(1) The exact numerical value of the correct does not always appear among the choices given. When this happens, select from among the choices the number that best approximates the exact numerical value.

(2) Unless otherwise specified, the domain of a function f is assumed to be the set of all real numbers x for which $f(x)$ is real number.

(3) The inverse of a trigonometric function f may be indicated using the inverse function notation f^{-1} or with the prefix "arc" (e.g, $\sin^{-1}x = \arcsin x$).

76. The function $f(x) = \begin{cases} x^2 & (x \leq 1) \\ e^x & (x > 1) \end{cases}$

(A) is continuous everywhere.

(B) is differentiable at $x = 1$.

(C) is continuous but not differentiable at $x = 1$.

(D) has a jump discontinuity at $x = 1$.

77. What is the horizontal asymptotes of $f(x) = \dfrac{2^{x+2} + 12}{2^{x+1} + 3}$?

(A) $y = 0,\ 2$

(B) $y = 2$

(C) $y = 4$

(D) $y = 2,\ 4$

78. $\dfrac{d}{dx}\left(5xe^{\ln x^4}\right)=$

(A) $25x^4$

(B) $20x^3$

(C) $5\ln x^4$

(D) $5+e^{\ln x^4}$

79. If $\dfrac{dy}{dx}=\sqrt{3y^2+1}$, what is $\dfrac{d^2y}{dx^2}$?

(A) $\dfrac{3y}{\sqrt{3y^2+1}}$

(B) $\dfrac{y}{\sqrt{3y^2+1}}$

(C) $3y$

(D) $\dfrac{3y}{3y^2+1}$

80. If $\displaystyle\int_a^b f(x)dx = 5$ and $\displaystyle\int_a^b g(x)dx = -6$, which of the following must be true?

I. $\displaystyle\int_a^b (f(x) - g(x))dx = 11$

II. $\displaystyle\int_a^b f(x)g(x)dx = -30$

III. $f(x) > g(x)$ for $a \le x \le b$

(A) I only

(B) I and II only

(C) I and III only

(D) II and III only

81. A particle moves along the x-axis so that its velocity v at time $t \ge 0$ is given by $v(t) = 2 + \sin(e^t)$. At time $t = 0$, the particle is at $x = -3$. What is the position of the particle at time $t = 3$?

(A) -2.72

(B) 0

(C) 3.61

(D) 6.61

82. Consider the differential equation $\frac{dy}{dx} = x(y^2 + 1)$. What is the particular solution $y = f(x)$ to the given differential equation with the initial condition $f(0) = 0$?

(A) $y = \tan\left(\frac{1}{2}x^2\right)$

(B) $y = \frac{1}{2}x^2 + x$

(C) $y = \frac{1}{3}x^3 + x$

(D) $y = \tan^{-1}\left(\frac{1}{2}x^2\right)$

83. An oil tank holds 800 gallons of oil at time $t = 0$. During the time interval $0 \le t \le 10$ hours, oil pumped into the tank at the rate $P(t) = 10\sqrt{t}\,\sin\left(\frac{t}{3}\right)$ gallons per hour. During the same time interval, oil is removed from the tank at the rate $R(t) = 8\sin^2\left(\frac{t}{5}\right)$ gallons per hour. How many gallons of oil are in the tank at time $t = 10$?

(A) 719
(B) 786
(C) 832
(D) 877

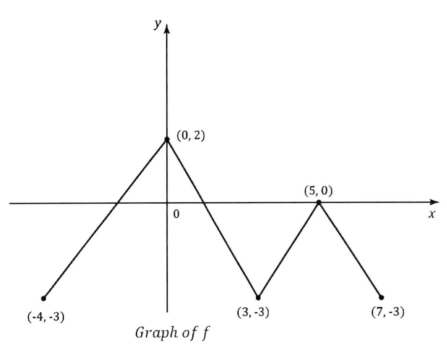

Graph of f

84. The graph of the function f is shown above. Let g be the function given by $g(x) = 2x + \int_0^x f(t)dt$. Which of the following must be true?

(A) g has relative minimum value at $x = -2$.

(B) g has inflection point at $x = -2,\ 2$.

(C) g is concave up at $x = \dfrac{19}{3}$.

(D) g has relative maximum at $x = \dfrac{12}{5},\ \dfrac{19}{3}$.

85. If $\int \{f(x)+3\}dx = \frac{1}{4}x^4 - x^3 - \frac{9}{2}x^2 + 4x + C$, what is the relative maximum of $y = f(x)$?

(A) -1

(B) 0

(C) 4

(D) 6

86. Let $x^2 + 4x + 1$ (ft^2) denote the area of the water surface when the water depth is x (ft) in a jar. What is the volume of water when the water depth is 3 ft?

(A) 20ft^3

(B) 24ft^3

(C) 30ft^3

(D) 36ft^3

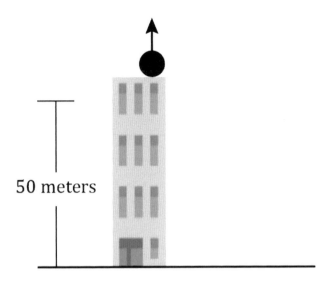

50 meters

87. As shown in the picture above, the velocity $v(t)$ at time t of the ball thrown straight up at a speed of $30\,\text{meter/sec}$ from the roof of a 50m-high building is called $v(t) = 30 - 6t$. What is the height from the ground when this ball reaches its maximum height?
(A) $75\,\text{meter}$
(B) $100\,\text{meter}$
(C) $125\,\text{meter}$
(D) $150\,\text{meter}$

88. Let f be a function defined by $f(x) = \begin{cases} 1 + \sin x & \text{for } x \leq 0 \\ e^{2x} & \text{for } x > 0 \end{cases}$.

Which of the following must be true?

(A) Function f is discontinuous at $x = 0$.

(B) Function f is differentiable at $x = 0$.

(C) The average value of f on the interval $[-\frac{1}{2}, \frac{1}{2}]$ is 4.947.

(D) None of these are true.

89. For $0 \le t \le 10$, a particle is moving along the $x-$axis. The particle's position, $x(t)$, is not explicitly given. The velocity of the particle is given by $v(t) = 5\sin(e^{\frac{t}{3}}) + 2$. The position of the particle is given by $x(0) = 11$. Which of the following must be true?

(A) The average velocity of the particle for the time period $0 \le t \le 10$ is 4.25.

(B) The total distance traveled by the particle from time $t = 0$ to $t = 10$ is 24.25.

(C) The speed of the particle is decreasing at time $t = 5$.

(D) For $0 \le t \le 10$, the particle changes direction six times.

90. $\frac{d}{dx}\{f(x) + g(x)\} = 3$, $\frac{d}{dx}\{f(x) \cdot g(x)\} = 4x + 1$. $f(0) = 2$, $g(0) = 1$. Which of the following expresses $f(x) + g(x) + f(x) \cdot g(x)$ correctly?

(A) $4x + 4$

(B) $2x^2 + 8x + 5$

(C) $2x^2 + x + 5$

(D) $2x^2 + 4x + 5$

AP Calculus AB Practice Test 1
❯ Answer Key

Part A

1. (C)	6. (D)	11. (D)	16. (B)	21. (A)	26. (D)
2. (A)	7. (A)	12. (C)	17. (D)	22. (C)	27. (A)
3. (A)	8. (D)	13. (C)	18. (C)	23. (A)	28. (A)
4. (D)	9. (C)	14. (C)	19. (A)	24. (D)	29. (C)
5. (B)	10. (A)	15. (B)	20. (C)	25. (A)	30. (B)

Part B

76. (D)	79. (C)	82. (A)	85. (D)	88. (C)
77. (D)	80. (A)	83. (D)	86. (C)	89. (C)
78. (A)	81. (C)	84. (D)	87. (C)	90. (D)

Part A

1. (C)

$x^4 = u \implies$ From $4x^3 = \dfrac{du}{dx}$, $\displaystyle\int x^3 \cos u \dfrac{du}{4x^3} = \dfrac{1}{4}\int \cos u\, du.$

$\implies \dfrac{1}{4}\sin u + C \implies \dfrac{1}{4}\sin(x^4) + C$

2. (A)

Let $e^{t^5} = f(t)$.

$\displaystyle\lim_{x \to 1}\dfrac{F(x) - F(1)}{3x^3 - 3}$ is the form of $\dfrac{0}{0}$, so we can use L'Hopital's Rule!

$\displaystyle\lim_{x \to 1}\dfrac{f(x)}{9x^2} = \dfrac{f(1)}{9} = \dfrac{e}{9}$

3. (A)

$3(xy + y^2)^2 \cdot (x'y + xy' + 2yy') = 8$

$\implies 3(1+1)^2 \cdot (1 + \dfrac{dy}{dx} + 2\dfrac{dy}{dx}) = 8$

$\implies 12(1 + 3\dfrac{dy}{dx}) = 8 \implies 1 + 3\dfrac{dy}{dx} = \dfrac{2}{3}$

$\implies 3\dfrac{dy}{dx} = -\dfrac{1}{3}$

$\therefore \dfrac{dy}{dx} = -\dfrac{1}{9}$

4. (D)

$g \circ f \Rightarrow \{g(f(x))\}' \Rightarrow g'(f(x)) CDOT f'(x)$

Since $g'(x) = -\sin x$ and $f'(x) = 6x$,

$\Rightarrow -6x \sin(3x^2 + 1)$

5. (B)

Let $5x = u$, then $5 = \dfrac{du}{dx}$

$\dfrac{1}{5} \displaystyle\int_5^{10} f(u)\,du = \dfrac{1}{5}(F(10) - F(5))$

6. (D)

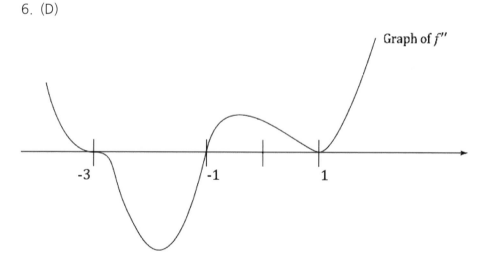

Graph of f''

f'' does sign change in $x = -3$ and $x = -1$, so f has inflection point at $x = -3, -1$.

7. (A)

We have to find inflection point, so we have to find the value of x that makes y'' sign change.

$$y' = 3x^2 + 12x \quad \Rightarrow \quad y'' = 6x + 12$$

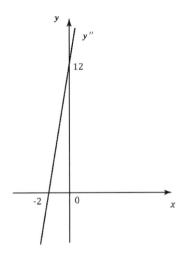

Therefore, $x = -2$

8. (D)

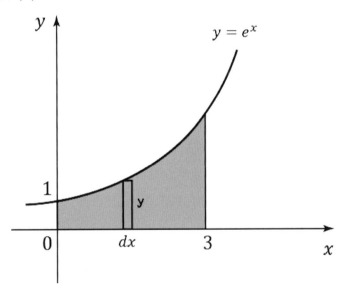

Since small Height is dx, therefore the length of square side is y.

$$\text{Therefore, Volume} = \int_0^3 \boxed{}_y \, y \, dx = \int_0^3 y^2 dx = \int_0^3 e^{2x} dx$$

Let $2x = u$, then from $2 = \dfrac{du}{dx}$, Volume $= \dfrac{1}{2}\int_0^6 e^u du = \dfrac{1}{2}(e^6 - 1)$

9. (C)
Graph like this situation could be drawn.

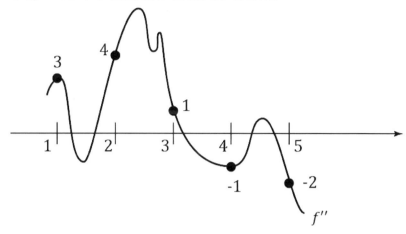

(A) Relative maximum should be where f' changes from positive to negative, or $f'(x)=0$ and $f''(x)<0$.

(B) f is concave up within the interval where $f''(x)>0$.

(C) Since in interval $(3, 4)$, f'' undergoes at least 1 sign change, therefore in this interval, function f gets at least 1 inflection point.

(D) Within interval where $f'(x)<0$, f is decreasing. We can't know whether f decreases or not within interval $(4, 5)$ only with the given information.

10. (A)

$$g'(3)=(f^{-1})'(3)=\frac{1}{f'(2)}=\frac{1}{5}$$

11. (D)

From $\int 1dy = \int x\cos x^2 dx$, let $x^2 = u$, then $2x = \dfrac{du}{dx}$

$y = \int x \cdot \cos u \cdot \dfrac{1}{2x}du$, so $y = \dfrac{1}{2}\sin x^2 + C$

Therefore, from $1 = \dfrac{1}{2}\sin 0 + C$, $C=1$, so $y = \dfrac{1}{2}\sin x^2 + 1$

12. (C)

We can infer the graph of f from f' graph, as follows.

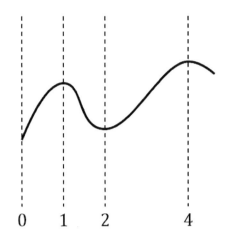

$$\textit{Graph of } f$$

I. $f(4) - f(1) = \int_1^4 f'(x) > 0 \ \Rightarrow \ f(4) > f(1)$

II. $f(2) - f(0) = \int_0^2 f'(x)dx > 0 \ \Rightarrow \ f(2) > f(0)$

III. From the f graph above, we can know that $f(1) > f(2)$

13. (C)

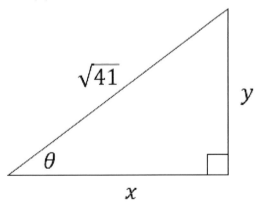

① $x^2 + y^2 = 41$, so when $y = 4$, $x = 5$.

② $\sin\theta = \dfrac{y}{\sqrt{41}}$, so $(\cos\theta)\dfrac{d\theta}{dt} = \dfrac{1}{\sqrt{41}}\dfrac{dy}{dt}$, and $\cos\theta = \dfrac{5}{\sqrt{41}}$.

③ $\dfrac{5}{\sqrt{41}} \times 5 = \dfrac{1}{\sqrt{41}}\dfrac{dy}{dt}$, therefore $\dfrac{dy}{dt} = 25$

14. (C)

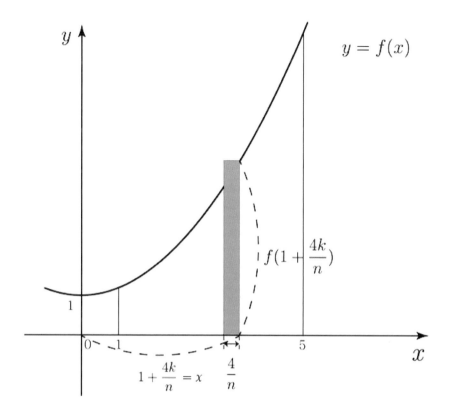

$\Rightarrow \dfrac{4}{n} = dx \qquad 1 + \dfrac{4}{n}k = x$

$\Rightarrow \displaystyle\lim_{n\to\infty}\sum_{k=1}^{n} f(1 + \dfrac{4}{n}k) \cdot \dfrac{4}{n}$, and since $f(x) = 3^x$, $\displaystyle\lim_{n\to\infty}\sum_{k=1}^{n} 3^{1+\frac{4}{n}k} \cdot \dfrac{4}{n}$

15. (B)

$g(1) = f(1) = 3$, and $g'(x) = 3x^2 \cdot f(x) + x^3 f'(x)$,

so $g'(1) = 3f(1) + f'(1) = 3 \cdot 3 + 2 = 11$

Slope is 11 and it passes $(1, 3)$, so from $y - 3 = 11(x - 1)$,

$g(x) = 11x - 8$

16. (B)

$\lim\limits_{x \to 5} \dfrac{2f(x)}{(x-5)^3}$ is form of $\dfrac{0}{0}$, so use L'Hopital's Rule!

$\Rightarrow \lim\limits_{x \to 5} \dfrac{2f'(x)}{3(x-5)^2}$ is also form of $\dfrac{0}{0}$ $\Rightarrow \lim\limits_{x \to 5} \dfrac{2f''(x)}{6(x-5)}$ is also form of $\dfrac{0}{0}$

$\Rightarrow \lim\limits_{x \to 5} \dfrac{2f'''(x)}{6} = \dfrac{4}{6} = \dfrac{2}{3}$

17. (D)

(A) By Intermediate Value Theorem (IVT), at least one c exists that satisfies $f(c) = 0$ in $[1, 5]$.

(B) f is continuous function, so by Extreme Value Theorem, Maximum Value or Minimum Value exists.

(C) f is differentiable in the given interval, so by Mean Value Theorem (MVT), at least one c exists that satisfies $f'(c) = \dfrac{3-(-3)}{5-1} = \dfrac{6}{4} = \dfrac{3}{2}$.

(D) Rolle's Theorem can be applied if f is differentiable and $f(1) = f(5)$, but since $f(1) \ne f(5)$, we cannot use Rolle's Theorem.

18. (C)

Since $g(x) = F(x) - F(-3)$, $g'(x) = f(x)$

That is, $f(x)$ is the graph of $g'(x)$.

(A) Since $g'(1)$, $g'(4)$, $g'(5)$ exist, g is differentiable at $x = 1, 4, 5$.

(B) g' has local minimum at the point where it changes from negative to positive, so g has local minimum at $x = -2$ and $x = 2$.

(C) Since g' is decreasing at $0 \le x \le 1$, g is concave down, and since $g' < 0$, g is decreasing.

(D) g is increasing at $2 \le x \le 6$, and concave up at $1 \le x \le 4$, $5 \le x \le 6$.

19. (A)

$h(x) = F(3x) - F(1)$, so $h'(x) = 3f(3x)$, therefore $h'(1) = 3f(3) = 3$.

20. (C)

(A) Use Mean Value Theorem.

If $H''(t) > 0$, graph of $H(t)$ is increasing and concave up.

Therefore, $H'(1.5) < H'(2.5) < H'(3.5) < H'(4.5)$ should be held.

① $H'(t) = \dfrac{22 - 17}{2 - 1} = 5 \ (1 < t < 2)$

② $H'(t) = \dfrac{36 - 22}{3 - 2} = 14 \ (2 < t < 3)$

③ $H'(t) = \dfrac{45 - 36}{4 - 3} = 9 \ (3 < t < 4)$

④ $H'(t) = \dfrac{78 - 45}{5 - 4} = 33 \ (4 < t < 5)$

Therefore, $H''(t) > 0$ is not held.

(B) $\displaystyle\int_1^5 H'(t)dt = H(5) - H(1) = 78 - 17 = 61$

(C) $\dfrac{\displaystyle\int_1^5 H(t)dt}{5 - 1} = \dfrac{1}{4}\displaystyle\int_1^5 H(t)dt$

$\displaystyle\int_1^5 H(t)dt = 17 + 22 + 36 + 45 = 120$, so $\dfrac{1}{4}\displaystyle\int_1^5 H(t)dt = 30$

(D) We can't know the value of t that has inflection point only with the information given.

21. (A)

• By $\displaystyle\lim_{x \to 0-} \dfrac{3}{1 + 5^{-\infty}} = \lim_{x \to 0-} \dfrac{3}{1 + \dfrac{1}{5^{\infty}}}$, since $\dfrac{1}{5^{\infty}} = 0$, $\displaystyle\lim_{x \to 0} \dfrac{3}{1 + 5^{\frac{1}{x}}} = 3$

• $\displaystyle\lim_{x \to 0+} \dfrac{3}{1 + 5^{\infty}} = 0$

• $\displaystyle\lim_{x \to 0-} \dfrac{3}{1 + 5^{\frac{1}{x}}} \neq \lim_{x \to 0+} \dfrac{3}{1 + 5^{\frac{1}{x}}}$, so Nonexistent.

22. (C)

(A) Change direction is when sign of velocity is changed, so when $t = 2, 5$, this object changes direction.

(B) Since $\text{Speed} = |\,Velocity\,|$, the graph of speed is as follows.

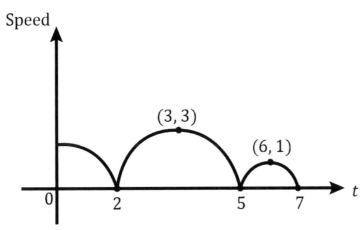

Therefore, the maximum speed is when time $t = 3$.

(C) Inflection point is when velocity graph changes from increasing to decreasing, or decreasing to increasing.

(D) Since the sign of velocity is negative and $V'(t) = a(t) > 0$ at $0 \leq t \leq 2$, speed is decreasing.

23. (A)

$\sec^2(x^2 y) \cdot (x^2 y)' = x'$

$\Rightarrow \sec^2(x^2 y) \cdot (2xy + x^2 y') = 1$

$\Rightarrow 2xy \cdot \sec^2(x^2 y) + x^2 \cdot \sec^2(x^2 y) \cdot y' = 1$

$\Rightarrow y' = \dfrac{1}{x^2 \cdot \sec^2(x^2 y)} - \dfrac{2xy\,\sec^2(x^2 y)}{x^2 \sec^2(x^2 y)}$

$\Rightarrow y' = \dfrac{\cos^2(x^2 y) - 2xy}{x^2}$

24. (D)

$$\int \frac{1}{x^2 + 10x + 25 + 1} dx = \int \frac{1}{1 + (x+5)^2} dx$$

Let $x + 5 = u$, then $\int \frac{1}{1 + u^2} du = \tan^{-1}u + C$

Therefore, $\tan^{-1}(x+5) + C$

25. (A)

$$2xx' + 2yy' = 0 \Rightarrow y' = -\frac{x}{y} \Rightarrow \frac{dy}{dx} = -\frac{x}{y}$$

$$\Rightarrow \frac{d}{dx}\frac{dy}{dx} = \frac{d}{dx}(-\frac{x}{y}) = -\frac{x'y - xy'}{y^2} = -\frac{y - x\frac{dy}{dx}}{y^2}$$

Since $\frac{dy}{dx} = -\frac{x}{y}$, $\frac{d^2y}{dx^2} = -\frac{y + \frac{x^2}{y}}{y^2}$

Therefore, put $x = y = 1$, then the answer is -2.

26. (D)

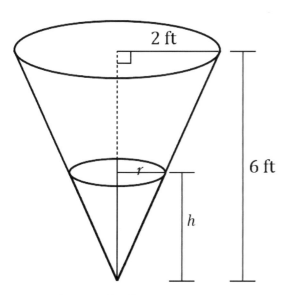

$2 : r = 6 : h$

$\implies r = \frac{1}{3}h$

① Volume $V = \frac{1}{3}\pi r^2 h$, $r = \frac{1}{3}h$,

so $V = \frac{1}{3}\pi \times \frac{1}{9}h^3$, so $V = \frac{1}{27}\pi h^3$

② $\frac{dV}{dt} = \frac{1}{9}\pi h^2 \frac{dh}{dt}$,

since $h = \frac{2}{3}$ ft and $\frac{dh}{dt} = -\frac{1}{10}$ ft/sec,

Therefore,

$\frac{dV}{dt} = \frac{1}{9}\pi \times \frac{4}{9}$ ft$^2 \times (-\frac{1}{10}$ ft/sec$)$,

$\frac{dV}{dt} = -\frac{2\pi}{405}$ ft^3/sec

However, what we should obtain is the rate water is leaking, so the answer is

(D) $\frac{2\pi}{405}$ ft^3/sec.

27. (A)

From $\lim\limits_{h\to 0}\dfrac{f(x+h)-f(x)}{h}=f'(x)$, $\lim\limits_{h\to 0}\dfrac{f(1+h)-f(1)}{h}=f'(1)=3$, so function f is differentiable at

$x=1$, so $\lim\limits_{x\to 1}f(x)$ and $f(1)$ exist.

Therefore, I is true.

$f'(1)=3$ means that slope of line tangent at $x=1$ is 3. We can't know if $f(1)=3$ or

$\lim\limits_{x\to 1}f(x)=3$.

Let's look at the following example.

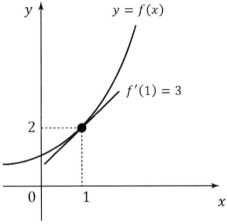

That is, the tangent line slope at point $(1,2)$ on $y=f(x)$ is 3, but either $f(1)=2$ or

$\lim\limits_{x\to 1-}f(x)=\lim\limits_{x\to 1+}f(x)=2$ can be established, so II and III both cannot always be true.

28. (A)

This is the $y-$axis rotation, and since the choices are all about dx, we can use Shell Method.

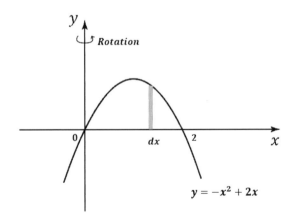

$V=2\pi\displaystyle\int_{0}^{2}xydx$, and $y=-x^2+2x$,

therefore $V=2\pi\displaystyle\int_{0}^{2}x(-x^2+2x)dx=2\pi\displaystyle\int_{0}^{2}(-x^3+2x^2)dx$.

29. (C)

(A) The following case may occur.

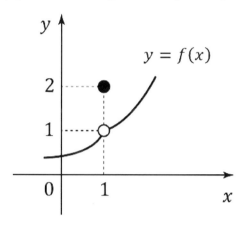

⇒ Although $\lim\limits_{x\to 1-} f(x) = \lim\limits_{x\to 1+} f(x)$,

$\lim\limits_{x\to 1} f(x) \neq f(1)$.

(B) The following case may occur.

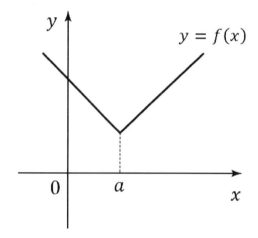

⇒ Although $y = f(x)$ is continuous at $x = a$,
it is not differentiable.

(C) Since $\lim\limits_{x\to 0} \dfrac{f(x) - f(0)}{x - 0} = f'(0) = 3$, function f is differentiable at $x = 0$. Therefore, function f is

continuous at $x = 0$ and $\lim\limits_{x\to 0} f(x)$ exists. Also, $f(0)$ exists.

(D) The following case may occur.

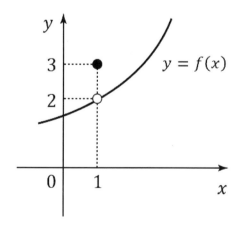

⇒ Although $f(1) = 3$,
$\lim\limits_{x\to 1-} f(x) = \lim\limits_{x\to 1+} f(x) = 2$.

30. (B)

$f(2)$ and $f(8)$ are relative minimum, so they are candidates of absolute minimum.

$f(8) - f(2) = \int_2^8 f'(x)dx = 7 + 1 - 5 = 3 > 0$, therefore $f(8) > f(2)$.

Therefore, Absolute Minimum Value is $f(2)$.

$f(9) - f(2) = \int_2^9 f'(x)dx = 7 + 1 - 5 + 1 = 4$, and $f(9) = 5$, so $f(2) = 1$.

Part B

76. (D)

$$\lim_{x \to 1-} x^2 = 1, \quad \lim_{x \to 1+} e^x = e, \quad f(1) = 1$$

Therefore, function f has discontinuity at $x = 1$, and is not differentiable.

Since $\lim_{x \to 1-} f(x) \neq \lim_{x \to 1+} f(x)$, so function f has jump discontinuity at $x = 1$.

77. (D)

$$\lim_{x \to \infty} \frac{2^2 \times 2^x + 12}{2 \times 2^x + 3} \Rightarrow \lim_{x \to \infty} \frac{2^2 + \dfrac{12}{2^x}}{2 + \dfrac{3}{2^x}} \Rightarrow \lim_{x \to \infty} \frac{2^2 + \dfrac{12}{\infty}}{2 + \dfrac{3}{\infty}} = 2$$

$$\lim_{x \to -\infty} \frac{2^2 \times 2^x + 12}{2 \times 2^x + 3} \Rightarrow \frac{\dfrac{4}{2^\infty} + 12}{\dfrac{2}{2^\infty} + 3} = 4$$

Therefore, horizontal asymptote is $y = 2,\ 4$

78. (A)

$e^{\ln x^4} = (x^4)^{\ln e} = x^4$, so $5xe^{\ln x^4} = 5x^5$

Therefore, $\dfrac{d}{dx}(5x^5) = 25x^4$

79. (C)

$$\frac{d^2y}{dx^2} = \frac{d}{dx}\frac{dy}{dx} = \frac{d}{dx}(\sqrt{3y^2+1}) \implies \frac{dy}{dx}\frac{d}{dy}(3y^2+1)^{\frac{1}{2}}$$

$$\implies \frac{d}{dy}(3y^2+1)^{\frac{1}{2}} = \frac{1}{2}(3y^2+1)^{-\frac{1}{2}} \cdot 6y = \frac{6y}{2\sqrt{3y^2+1}}$$

Therefore, from $\frac{d^2y}{dx^2} = \frac{dy}{dx}(\frac{6y}{2\sqrt{3y^2+1}})$, $\frac{dy}{dx} = \sqrt{3y^2+1}$.

$$\frac{d^2y}{dx^2} = \sqrt{3y^2+1} \cdot \frac{6y}{2\sqrt{3y^2+1}} = 3y$$

80. (A)

I. $\displaystyle\int_a^b (f(x)-g(x))dx = \int_a^b f(x)dx - \int_a^b g(x)dx = 5+6 = 11$

II. $\displaystyle\int_a^b f(x)g(x)dx \neq \int_a^b f(x)dx \cdot \int_a^b g(x)dx$

III. Situations like these can happen.

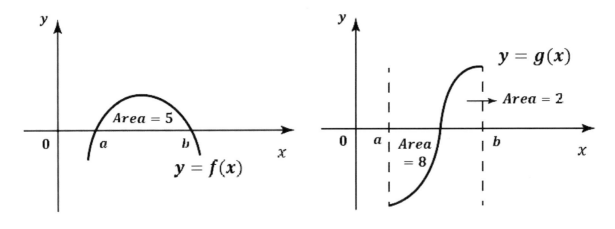

Therefore, we can't surely say that $f(x) > g(x)$.

81. (C)

Let time as t and position as x then $x(3) - x(0) = \int_0^3 v(t)dt$

$x(3) = x(0) + \int_0^3 (2 + \sin(e^t))dt \approx 3.61$

82. (A)

From $\int \dfrac{1}{y^2 + 1}dy = \int x\,dx$, $\tan^{-1}y = \dfrac{1}{2}x^2 + C$

Since $\tan^{-1}(0) = C$, $C = 0$.

Therefore, by $\tan^{-1}y = \dfrac{1}{2}x^2$, $y = \tan\left(\dfrac{1}{2}x^2\right)$

83. (D)

Let $P(t)$ as $f'(t)$, $R(t)$ as $g'(t)$ and the change rate of oil amount in the tank as $h'(t)$, then $h'(t) = f'(t) - g'(t)$.

Let the amount of oil in the tank at time $t = 10$ as $h(10)$, then

$h(10) - h(0) = \int_0^{10} h'(t)dt$

therefore, $h(10) = 800 + \int_0^{10} h'(t)dt \approx 877$

84. (D)

$$\int_0^x f(t)dt = F(x) - F(0), \quad g(x) = 2x + F(x) - F(0), \text{ so } g'(x) = 2 + f(x)$$

That is, graph of $g'(x)$ is a graph in which the graph of $f(x)$ is shifted by 2 on the y-axis. $g'(x)$ graph is like below.

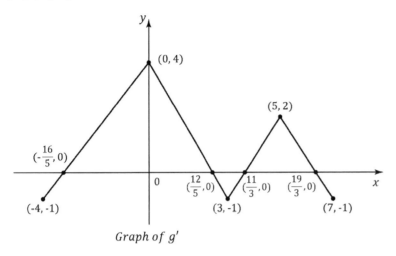

Graph of g'

(A) g has relative minimum at $x = -\dfrac{16}{5}, \ \dfrac{11}{3}$.

(B) g has inflection point at $x = 0, \ 3, \ 5$.

(C) g is concave down at $x = \dfrac{19}{3}$.

(D) g has relative maximum at $x = \dfrac{12}{5}, \ \dfrac{19}{3}$.

85. (D)

Differentiate both sides by x

$f(x) + 3 = x^3 - 3x^2 - 9x + 4$, therefore $f(x) = x^3 - 3x^2 - 9x + 1$

Since $f'(x) = 3x^2 - 6x - 9 = 0$, $f'(x) = 3(x^2 - 2x - 3) = 0$, and $x = -1, \ 3$.

Shape of graph of f is as follows, approximately.

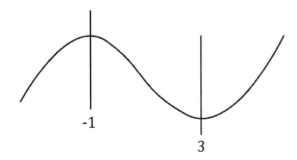

Therefore, $f(-1) = -1 - 3 + 9 + 1 = 6$

86. (C)

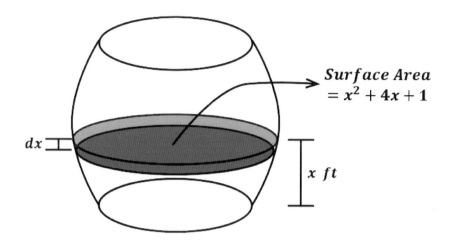

Surface Area
$= x^2 + 4x + 1$

dx

$x \, ft$

Let the volume of water as V.

$\Rightarrow V = \int_0^3 (\text{Surface Area}) \times (\text{Small Height})$

$\Rightarrow V = \int_0^3 (x^2 + 4x + 1) dx$

$\Rightarrow V = [\frac{1}{3}x^3 + 2x^2 + x]_0^3$

$V = \frac{1}{3} \times 3^3 + 2 \times 3^2 + 3 = 30 \text{ft}^3$

87. (C)

Velocity when it reaches to the maximum height becomes 0.

Therefore, the time when it reaches to the maximum height would be $t = 5$.

Let the distance between the roof of the building and the maximum height of the ball as $h(t)$,

then $h(t) = \int_0^5 (30 - 6t) dt = [30t - 3t^2]_0^5 = 150 - 75 = 75 \text{meter}$.

Height from the base ground would be 125 meter.

or

Let the height of the ball over time t as $h(t)$, then $h(0) = 50$.

Therefore, $h(5) - h(0) = \int_0^5 v(t) dt \Rightarrow h(5) = 125$

88. (C)

(A) $\lim_{x \to 0-} (1 + \sin x) = 1$, $\lim_{x \to 0+} (e^{2x}) = 1$, $f(0) = 1$,

and $\lim_{x \to 0} f(x) = f(0)$ is satisfied, so function f is continuous at $x = 0$.

(B) $f'(x) = (1 + \sin x)' = \cos x \Rightarrow f'(0) = 1$

$f'(x) = (e^{2x})' = 2e^{2x} \Rightarrow f'(0) = 2$

Therefore, function f is not differentiable at $x = 0$.

(C) The average value of f on the interval $[-\frac{1}{2}, \frac{1}{2}]$ is $\dfrac{\displaystyle\int_{-\frac{1}{2}}^{\frac{1}{2}} f(x)dx}{\dfrac{1}{4}} = 4\displaystyle\int_{-\frac{1}{2}}^{\frac{1}{2}} f(x)dx$.

$\displaystyle\int_{-\frac{1}{2}}^{\frac{1}{2}} f(x)dx = \int_{-\frac{1}{2}}^{0} f(x)dx + \int_{0}^{\frac{1}{2}} f(x)dx$, and $\displaystyle\int_{-\frac{1}{2}}^{0} (1 + \sin x)dx + \int_{0}^{\frac{1}{2}} e^{2x} dx \approx 1.237$

Therefore, average value is $4 \times 1.237 \approx 4.947$.

89. (C)

(A) $\dfrac{1}{10} \displaystyle\int_{0}^{10} V(t)dt \approx 2.99$

(B) Total distance $= \displaystyle\int_{0}^{10} |5\sin(e^{\frac{t}{3}}) + 2| dt \approx 40.92$

(C) $V(5) \approx -2.18$

$a(5) \approx 4.85$

When time $t = 5$, sign of velocity and that of acceleration is different, so the speed is decreasing.

(D) Using calculator, let's draw graph of velocity. The sign of velocity is changed 8 times. That is, change direction happened 8 times.

90. (D)

$f(x) + g(x) = \displaystyle\int 3dx = 3x + C_1$, $f(x) \cdot g(x) = 2x^2 + x + C_2$

Substitute $x = 0$, then $f(0) + g(0) = C_1$, $f(0) \cdot g(0) = C_2$,

therefore $C_1 = 3$, $C_2 = 2$. Therefore, $f(x) + g(x) = 3x + 3$, $f(x) \cdot g(x) = 2x^2 + x + 2$.

$\Rightarrow f(x) + g(x) + f(x) \cdot g(x) = 2x^2 + 4x + 5$.

AP Calculus AB

Practice Test 2

CALCULUS AB
SECTION I, Part A
Time-60 minutes
Number of questions-30

A CALCULATOR MAY NOT BE USED ON THIS PART OF THE EXAMINATION.

Directions: Solve each of the following problems, using the available space for scratchwork. After examining the form of the choices, decide which is the best of the choices given and fill in the corresponding oval on the answer sheet. No credit will be given for anything written in the test book. Do not spend too much time on any one problem.

In this test:

(1) Unless otherwise specified, the domain of a function is assumed to be the set of all real numbers x for which $f(x)$ is a real number.

(B) The inverse of a trigonometric function f may be indicated using the inverse function notation f^{-1} or with the prefix "arc" (e.g., $\sin^{-1} = \arcsin x$).

1. $\dfrac{d}{dx}\sin^2(x^2) =$

(A) $2\cos(x^2)$

(B) $2x\sin(2x^2)$

(C) $2\cos(x^2)\sin(x^2)$

(D) $2x\cos(x^2)\sin(x^2)$

2. If $f(x) = \ln(x + 3 + e^{2x})$, then $f'(0)$ is

(A) $\dfrac{3}{4}$

(B) $\dfrac{1}{2}$

(C) $\dfrac{1}{4}$

(D) 0

3. $\displaystyle\int_1^2 \frac{2x}{x^2+3}\,dx =$

(A) $\ln 2$

(B) $\dfrac{1}{2}\ln 2$

(C) $\ln\dfrac{7}{4}$

(D) $\dfrac{1}{2}\ln\dfrac{7}{4}$

4. The function f given by $f(x) = 4x^3 - 12x^2 + 6x$ has an inflection point at $x =$

(A) 0

(B) 1

(C) $\dfrac{3}{2}$

(D) 2

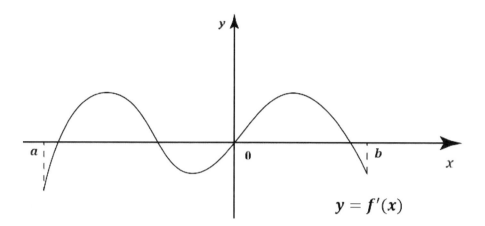

$$y = f'(x)$$

5. The graph of f', the derivative of f, is shown in the figure above. Which of the following describes f on the open interval (a, b)?
(A) Three inflection points and one relative minimum
(B) Two inflection points and two relative maxima
(C) Two relative maxima and one relative minimum
(D) Three inflection points and two relative minima

6. $\displaystyle\int_{2}^{4} \frac{x+1}{x^2}\,dx =$

(A) $\ln 4 - \dfrac{1}{2}$

(B) $\dfrac{1}{2} - \ln 2$

(C) $\ln 2$

(D) $\ln 2 + \dfrac{1}{4}$

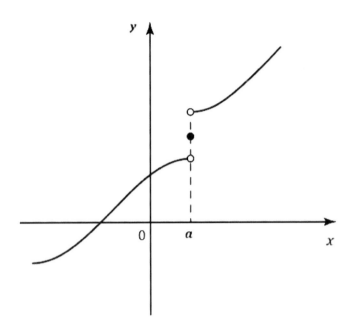

7. The graph of a function f is shown above. Which of the following statements about f is false?

(A) f is not differentiable at $x = a$.

(B) f is discontinuous at $x = a$.

(C) $\lim\limits_{x \to a} f(x)$ exist.

(D) $\lim\limits_{x \to a-} f(x) = \lim\limits_{x \to a+} f(x)$

t (sec)	0	1	2	3
$a(t)$ (ft/sec^2)	6	7	5	9

8. The data for the acceleration $a(t)$ of a particle from 0 to 3 seconds are given in the table above. If the velocity at $t=0$ is 10 feet per second, the approximate value of the velocity at $t=3$, computed using a right-hand Riemann Sum with three subintervals of equal length is
(A) 12 ft/sec
(B) 18 ft/sec
(C) 21 ft/sec
(D) 31 ft/sec

9. At what point on the graph of $y = 2x^2$ is the tangent line parallel to the line $4x - 4y = 2$?
(A) $(\frac{1}{4}, \frac{3}{4})$
(B) $(\frac{1}{4}, \frac{1}{8})$
(C) $(\frac{1}{4}, \frac{1}{4})$
(D) $(\frac{1}{4}, \frac{1}{16})$

10. If f is the function defined by $f(x) = 2x^4 - 4x^2$, what are all the $x-$coordinates of points of relative maximum for the graph of f?

(A) 0 only

(B) -1 and 0 only

(C) 0 and 1 only

(D) -1 and 1 only

11. $\lim\limits_{x \to 2} \dfrac{\displaystyle\int_{2}^{x} \sin(t^2 + t)dt}{x - 2} =$

(A) $\cos 6$

(B) $\sin 6$

(C) $\sec 6$

(D) $\csc 6$

12. If $\lim\limits_{x \to 1} f(x) = 2$, which of the following must be true?

I. $f(1) = 2$

II. f is continuous at $x = 1$

III. f is differentiable at $x = 1$

(A) None

(B) I only

(C) I and II only

(D) I and III only

13. The function f has a first derivative given by $f'(x) = x^2 - 6x + 8$. At what values of x does f is concave down?

(A) $x > 4$ or $x < 2$

(B) $2 < x < 4$

(C) $x > 3$

(D) $x < 3$

14. If $\begin{cases} f(x) = \dfrac{\sqrt{2x+9} - \sqrt{x+12}}{x-3} \\ f(3) = k \end{cases}$, and if f is continuous at $x = 3$, then $k =$

(A) 0

(B) 1

(C) $\dfrac{\sqrt{15}}{30}$

(D) $\dfrac{\sqrt{15}}{15}$

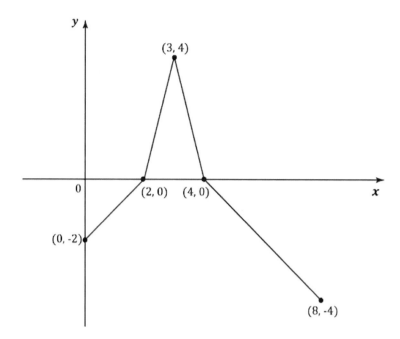

Graph of f'

15. For $0 \leq x \leq 8$, the graph of f', the derivative of f, is piecewise linear as shown above. If $f(3) = 2$, what is the maximum value of f on the interval?

(A) $\dfrac{1}{2}$

(B) 2

(C) $\dfrac{5}{2}$

(D) 4

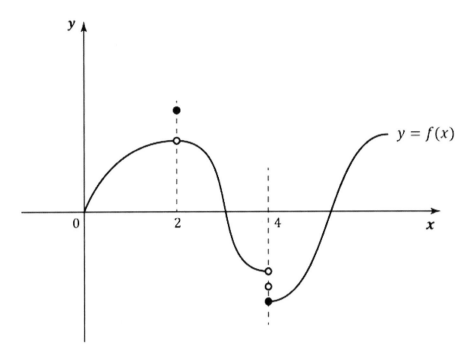

Graph of f

16. The figure above shows the graph of the function f. Which of the following statements are true?

I. f has a jump discontinuity at $x = 2$.

II. f has a removable discontinuity at $x = 4$.

III. $\lim_{x \to 2} f(x)$ exists.

(A) I only

(B) II only

(C) III only

(D) I and II only

17. The function f and g are given by $f(t) = \int_0^{2x} \sqrt{1+t^2}\,dt$ and $g(x) = f(e^{3x})$. Which of the following must be true?

(A) $f'(x) = \sqrt{1+4x^2}$

(B) $g'(x) = 2e^{3x}\sqrt{1+4e^{6x}}$

(C) The equation for the line tangent to the graph of $y = f(x)$ at $x = 0$ is $y = 2x$.

(D) None of above.

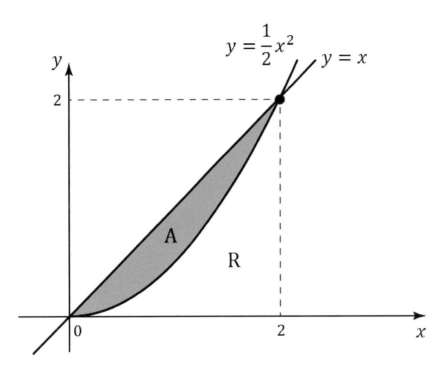

18. Let A be the region in the first quadrant enclosed by the graphs of $y = x$ and $y = \frac{1}{2}x^2$, as shown in the figure above. Which of the following must be true?

(A) The area of R is $R = \int_0^2 (\frac{1}{2}x^2 - x)dx$.

(B) The region A is the base of a solid, at each x the cross section parallel to the $y-$axis has area $R(x) = \cos(\frac{2\pi}{3}x)$. The volume of the solid is $\int_0^2 \cos(\frac{2\pi}{3}x)dy$.

(C) The volume of the solid generated when A is revolved about the horizontal line $y = -2$ is $\pi \int_0^2 (\frac{1}{4}x^4 - x^2)dx$.

(D) The volume of the solid generated when A is revolved about the $y-$axis is $2\pi \int_0^2 (x^2 - \frac{1}{2}x^3)dx$.

19. Let f be a twice differentiable function such that $f(2)=5$ and $f(4)=13$. Which of the following must be true for the function f on the interval $2 \le x \le 4$?

I. The average value of f' is 4.

II. The average rate of change of f is 4.

III. $f(c)=7$ for at least one c between 2 and 4.

(A) I only

(B) I and II only

(C) II and III only

(D) I, II, and III

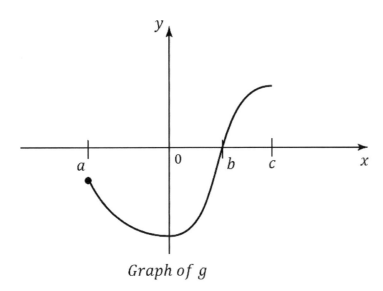

Graph of g

20. Let $f(x) = \int_a^x g(t)dt$, where g has the graph shown above. Which of the following could be the graph of f?

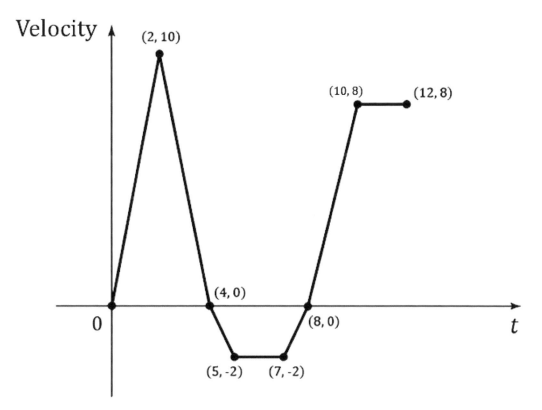

Velocity

(2, 10)

(10, 8)   (12, 8)

(4, 0)

0   (8, 0)

(5, -2)   (7, -2)

t

Graph of Velocity

21. An object moves along the $x-$axis. For $0 \le t \le 12$, the object's velocity is modeled by the piecewise-linear function defined by the graph above. The object is at position $x = 3$ when $t = 0$. Which of the following must be true?

(A) The object changes direction at $t = 5$ and $t = 7$.

(B) The total distance the object travels during the time interval $0 \le t \le 12$ is 38.

(C) The position of the object at $t = 4$ is 23.

(D) The acceleration of the object at $t = 2$ is 5.

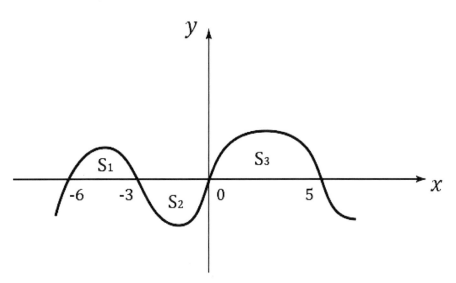

22. The graph of $y = f(x)$ is shown in the figure above. If S_1, S_2, and S_3 are positive numbers that represent the areas of the shaded regions, then in terms of S_1, S_2, and S_3,

$$3\int_{-6}^{-3} f(x)dx - 2\int_{-3}^{0} f(x)dx + 2\int_{0}^{5} f(x)dx =$$

(A) $3S_1 + 2S_2 + 2S_3$

(B) $3S_1 - 2S_2 + 2S_3$

(C) $-S_1 + 2S_2 - 3S_3$

(D) $3S_1 - 2S_2 - 2S_3$

23. A curve has slope $4x+2$ at each point (x, y) on the curve. Which of the following is an equation for this curve if it passes through the point $(-3, 4)$?

(A) $y = 2x^2 + 2x$

(B) $y = 4x$

(C) $y = 4x - 8$

(D) $y = 2x^2 + 2x - 8$

24. If $\dfrac{dy}{dx} = ay$ and a is a nonzero constant, then y could be

(A) $3e^{ax} + 2$

(B) $3e^{ax}$

(C) $ax + 1$

(D) ax

25. Let f be the function defined by $f(x) = \frac{1}{3}x^3 - 2x^2 + 3x + 1$. On which of the following intervals is the graph of f both increasing and concave up?

(A) $1 < x < 3$

(B) $x > 3$ or $x < 1$

(C) $0 < x < 1$

(D) $x > 3$

26. If the graph of $y = f(x)$ contains the point $(e, 1)$, $\dfrac{dy}{dx} = \dfrac{\ln x}{xy}$ and $f(x) > 0$ for $x > 1$ then $f(x) =$

(A) x

(B) $\sqrt{x}$

(C) $\ln x$

(D) $\sqrt{\ln x}$

27. The base of a solid is the region enclosed by the graph of $y = \ln x$, the coordinate axes, and the line $x = 2$ and $x = 5$. If all plane cross sections perpendicular to the x-axis are semi-circles, then its volume is

(A) $\dfrac{\pi}{2}\displaystyle\int_{2}^{5}(\ln x)^2 dx$

(B) $\dfrac{\pi}{8}\displaystyle\int_{2}^{5}(\ln x)^2 dx$

(C) $\pi\displaystyle\int_{2}^{5}(\ln x)dx$

(D) $\dfrac{\pi}{8}\displaystyle\int_{2}^{5}(\ln x)dx$

28. If $(xy + x^2)^3 = 8y$ and $x = y = 1$, then $\dfrac{dy}{dx}$ is

(A) -9
(B) -2
(C) 4
(D) 5

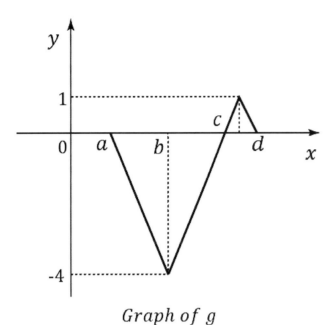

Graph of g

29. The graph of g is shown in the figure above. If $f(x) = \int_0^x g(t)dt$, for what value of x does $f(x)$ have an absolute maximum?

(A) a

(B) b

(C) c

(D) d

30. $\displaystyle\int \frac{1}{\sqrt{4-x^2}}dx =$

(A) $-\dfrac{2}{3}(4-x)^3 + C$

(B) $\sin^{-1}(\dfrac{1}{2}x) + C$

(C) $\dfrac{1}{2}\sin^{-1}(\dfrac{1}{2}x) + C$

(D) $\dfrac{1}{2}(4-x)^{\frac{3}{2}} + C$

CALCULUS AB
SECTION I, PART B
Time-45 minutes
Number of questions-15

A GRAPHING CALCULATOR IS REQUIRED FOR SOME QUESTIONS ON THIS PART OF THE EXAMINATION.

Directions: Solve each of the following problems, using the available space for scratch work. After examining the form of the choices, decide which is the best of choices given and fill in the corresponding oval on the answer sheet. No credit will be given for anything written in the test book. Do no spend too much time on any one problem.

BE SURE YOU ARE USING PAGE 3 OF THE ANSWER SHEET TO RECORD YOUR ANSWERS TO QUESTIONS NUMBERED 76-90.

YOU MAY NOT RETURN TO PAGE 2 OF THE ANSWER SHEET.

In this test:

(1) The exact numerical value of the correct does not always appear among the choices given. When this happens, select from among the choices the number that best approximates the exact numerical value.

(2) Unless otherwise specified, the domain of a function f is assumed to be the set of all real numbers x for which $f(x)$ is real number.

(3) The inverse of a trigonometric function f may be indicated using the inverse function notation f^{-1} or with the prefix "arc" (e.g, $\sin^{-1}x = \arcsin x$).

76. $\int \dfrac{e^{\tan x}}{\cos^2 x}\,dx =$

(A) $e^{\sin x} + C$

(B) $\sec^2 x \tan x + C$

(C) $\tan^2 x + C$

(D) $e^{\tan x} + C$

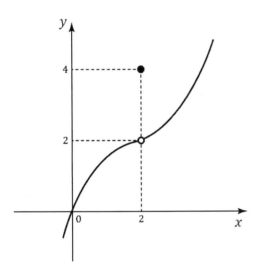

77. The graph of the function f is shown in the figure above. The value of $\lim\limits_{x \to 2} \tan(f(x))$ is

(A) -2.185

(B) -0.235

(C) 1.158

(D) 2.317

78. What are all values of x for which the function f defined by $f'(x)=(x-2)\sin x$ is both increasing and concave up?

(A) $-3.14<x<-1.8$ and $2<x<2.64$

(B) $0.85<x<0.31$

(C) $-3.14<x<-1.8$ and $0<x<2$

(D) $2<x<2.64$

79. Let $y=f(x)$ be the function defined by $f(x)=\sin(x^2+x)$, and let g be the function defined by $g(x)=-x^3-x^2-3$. The line tangent to the graph of f at $x=3$ is perpendicular to the graph of g at $x=k$, where k is a positive constant. What is the value of k?

(A) 0.056

(B) 0.083

(C) 0.102

(D) 0.143

80. The function f is continuous for $-1 \le x \le 3$ and differentiable for $-1 < x < 3$.
If $f(-1)=-2$ and $f(3)=6$, which of the following statements could be false?
(A) There exists, c, where $-1<c<3$, such that $f(c) \le f(x)$ for all x on the closed interval $-1 \le x \le 3$.
(B) There exists, c, where $-1<c<3$, such that $f(c)=0$.
(C) There exists, c, where $-1<c<3$, such that $f'(c)=2$.
(D) There exists, c, where $-1<c<3$, such that $f'(c)=0$.

81. A particle moves along the $x-$axis with velocity at time t given by $v(t)=\ln(t)CDOTsin(2t)$ for $0 \le t \le \pi$. What is the time t at which the particle is farthest to the right?
(A) 0
(B) 1
(C) 1.57
(D) 3.14

82. If $y = \sin(3x) + e^{5x}$, then $\dfrac{d^3 y}{dx^3} =$

(A) $-27\cos 3x + 125 e^{5x}$

(B) $-9\sin 3x + 25 e^{5x}$

(C) $27\sin 3x + 125 e^{5x}$

(D) $9\cos 3x + 25 e^{5x}$

83. A particle moves along the $x-$axis so that its velocity v at time t, for $0 \leq t \leq 8$, is given by $v(t) = \ln(t^2 - 4t + 5)$. The particle is at position $x = 10$ at time $t = 0$. Which of the following must be true?

(A) The acceleration of the particle at time $t = 3$ is 0.69.

(B) The average speed of the particle over the interval $0 \leq t \leq 8$ is 13.91.

(C) The position of the particle at time $t = 5$ is 4.84.

(D) The speed of the particle over the interval $0 < t < 1$ is decreasing.

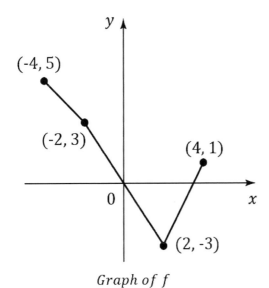

Graph of f

84. The graph of the function f above consists of three line segments. Let g be the function given by $g(x) = \int_{-4}^{2x} f(t)dt$. Which of the following must be true?

(A) $g'(2) = 1$

(B) $g''(1.5) = 8$

(C) $g(0) = 0$

(D) g has inflection point at $x = 2$.

85. For all values of x, the continuous function f is negative and increasing. Let g be the function given by $g(x) = \int_1^x f(t)\,dt$. Which of the following could be a table of values for g?

(A)

x	$g(x)$
1	10
2	5
3	3

(B)

x	$g(x)$
1	10
2	8
3	2

(C)

x	$g(x)$
1	3
2	5
3	10

(D)

x	$g(x)$
1	2
2	8
3	10

86. Let f be the function given by $f(x) = \int_1^x \sin\left(\frac{3}{2t^2}\right)dt$ for $1 \le x \le 3$. At which of the following values of x does f attain a local maximum?

(A) 1.45 only

(B) 2.05 only

(C) 1.45 and 2.51

(D) 2.05 and 2.90

x	1	3	5	7	9	11
$f(x)$	-1	2	3	0	1	-2

87. The table above shows selected values of a continuous function f. For $1 \le x \le 11$, what is the fewest possible number of times $f(x) = 0$?

(A) One

(B) Two

(C) Three

(D) Four

88. $\displaystyle\int_3^7 e^x \, dx =$

(A) $\displaystyle\lim_{n\to\infty}\sum_{k=1}^{n} e^{\left(3+\frac{4}{n}k\right)} \cdot \frac{4}{n}$

(B) $\displaystyle\lim_{n\to\infty}\sum_{k=1}^{n} e^{\frac{4}{n}k} \cdot \frac{4}{n}$

(C) $\displaystyle\lim_{n\to\infty}\sum_{k=1}^{n} e^{\left(3+\frac{4}{n}k\right)} \cdot \frac{1}{n}$

(D) $\displaystyle\lim_{n\to\infty}\sum_{k=1}^{n} e^{\frac{4}{n}k} \cdot \frac{1}{n}$

186

89. A man stands on the road 30 meters north of the crossing and watches an westbound car traveling at 10 meters per second. At how many meters per second is the car moving away from the man 5 seconds after it passes through the intersection? (A road track and a road cross at right angles)
(A) 6.88
(B) 7.26
(C) 8.57
(D) 9.13

x	$f(x)$	$f'(x)$
0	2	1
1	5	2
2	10	7

90. The table above gives selected values for a differentiable and increasing function f and its derivative. If g is the inverse function of f, what is the value of $g'(5)$?

(A) 1

(B) $\dfrac{1}{2}$

(C) $\dfrac{1}{7}$

(D) $\dfrac{1}{10}$

AP Calculus AB Practice Test 2
❯ Answer Key

Part A

1. (B)	6. (D)	11. (B)	16. (C)	21. (C)	26. (C)
2. (A)	7. (C)	12. (A)	17. (C)	22. (A)	27. (B)
3. (C)	8. (D)	13. (D)	18. (D)	23. (D)	28. (A)
4. (B)	9. (B)	14. (C)	19. (D)	24. (B)	29. (A)
5. (D)	10. (A)	15. (D)	20. (B)	25. (D)	30. (B)

Part B

76. (D)	79. (A)	82. (A)	85. (A)	88. (A)
77. (A)	80. (D)	83. (D)	86. (C)	89. (C)
78. (A)	81. (A)	84. (B)	87. (C)	90. (B)

Part A

1. (B)

$$\{(\sin(x^2))^2\}' = 2(\sin(x^2)) \cdot \cos(x^2) \cdot 2x$$
$$= 4x \cdot \sin(x^2) \cdot \cos(x^2)$$
$$= 2x \cdot 2\sin(x^2) \cdot \cos(x^2)$$
$$= 2x \sin(2x^2)$$

> ※ $\sin 2x = 2\sin x \cos x$

2. (A)

$$f'(x) = \frac{1 + 2e^x}{x + 3 + e^{2x}} \qquad \Rightarrow \qquad f'(0) = \frac{3}{4}$$

3. (C)

Let $x^2 + 3 = u$, then $2x = \dfrac{du}{dx}$.

Therefore, $\displaystyle\int_4^7 \frac{2x}{u} \cdot \frac{du}{2x} = \int_4^7 \frac{1}{u} du = [\ln u]_4^7$

$$= \ln 7 - \ln 4 = \ln \frac{7}{4}$$

4. (B)

$f'(x) = 12x^2 - 24x + 6 \implies f''(x) = 24x - 24$

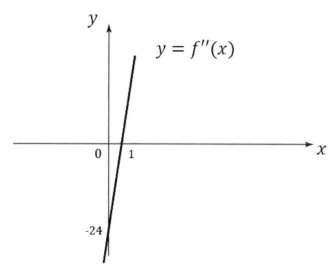

Like the graph above, inflection points is made at the x-value where sign change of f'' happens.

5. (D)

f has inflection point where graph of f' changes from increasing to decreasing or decreasing to incresaing. Therefore, f has three inflection points. Relative minimum is made where graph of f' changes from negative to positive, so f has two relative minima.

6. (D)

$$\int_2^4 (\frac{1}{x} + x^{-2})dx = [\ln x - \frac{1}{x}]_2^4$$

$$= \ln 4 - \frac{1}{4} - \ln 2 + \frac{1}{2} = \ln 2 + \frac{1}{4}$$

7. (C)

In the figure given, f is discontinuous at $x = 2$ so it becomes not differentiable.

$\lim\limits_{x \to a-} f(x) \neq \lim\limits_{x \to a+} f(x)$, and $\lim\limits_{x \to a} f(x)$ is nonexistent.

8. (D)

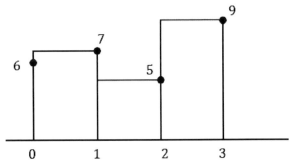

Area = $7 + 5 + 9 = 21$

Let velocity and acceleration at t be $V(t)$ and $a(t)$, respectively. Since $V(3) - V(0) = \int_0^3 a(t)dt$ and

$\int_0^3 a(t)dt = 21$, $V(3) = 10 + \int_0^3 a(t)dt = 31\,ft/\sec$

9. (B)

In $4x - 4y = 2$, $y = x - \dfrac{1}{2}$.

Slope of this is 1, so it should be like $y' = 4x = 1$. Therefore $x = \dfrac{1}{4}$.

$y = 2 \times (\dfrac{1}{4})^2 = \dfrac{1}{8}$.

10. (A)

From $f'(x) = 8x^3 - 8x = 0$, $x = -1,\ 0,\ 1$.

Since $f''(x) = 24x^2 - 8$, $f''(-1) > 0$, $f''(0) < 0$, $f''(1) > 0$.

Therefore, f has relative maximum at $x = 0$.

or

If we draw the graph of $f'(x) = 8x(x-1)(x+1)$,

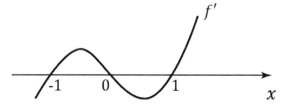

That is, at $x = 0$, f' changes from positive to negative, therefore at $x = 0$, function f has relative maximum.

11. (B)

Let $\sin(t^2 + t)$ as $f(t)$,

$$\lim_{x \to 2} \frac{\displaystyle\int_2^x f(t)dt}{x - 2} \quad \Rightarrow \quad \lim_{x \to 2} \frac{F(x) - F(2)}{x - 2},\ \text{and use L'Hopital's Rule, then}$$

$$\lim_{x \to 2} \frac{f(x)}{1} \quad \Rightarrow \quad \lim_{x \to 2}(\sin(x^2 + x)) = \sin 6.$$

12. (A)

We can't certify about continuity and differentiability only because $\lim_{x \to 1} f(x)$ exists. For example, case like below can be made.

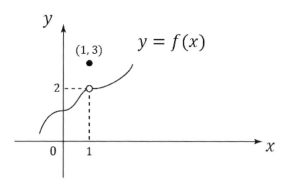

13. (D)

Function f becomes concave down where graph of f' is decreasing.

Graph of f' is like below.

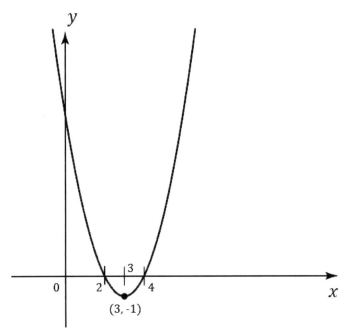

Therefore, x values that f is concave down are which is included in $x < 3$.

or

Within the interval where $f''(x) < 0$, f is concave down,

therefore from $f''(x) = 2x - 6 < 0$, $x < 3$.

14. (C)

By $\lim\limits_{x \to 3} \dfrac{\sqrt{2x+9} - \sqrt{x+12}}{x-3} = f(3) = k$,

$\lim\limits_{x \to 3} \dfrac{(x-3)}{(x-3)(\sqrt{2x+9} + \sqrt{x+12})} \Rightarrow \lim\limits_{x \to 3} \dfrac{1}{\sqrt{2x+9} + \sqrt{x+12}}$, therefore

$\dfrac{1}{2\sqrt{15}} = f(3) = k$. Therefore $k = \dfrac{\sqrt{15}}{30}$.

15. (D)

We can infer the graph of f from the graph of f' as follows.

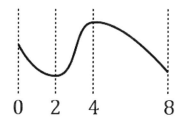

Graph of f

Values that can be absolute maximum are those when $x=0$ and $x=4$.

Therefore, $f(4)-f(0)=\int_0^4 f'(x)dx>0$, so $f(4)>f(0)$.

Therefore, Absolute maximum is $f(4)$.

$f(4)-f(3)=\int_3^4 f'(x)dx$, so $\int_3^4 f'(x)dx=\frac{1}{2}\times 4=2$.

Therefore, $f(4)=f(3)+2=4$.

16. (C)

I. f has removable discontinuity at $x=2$. ($\ast \lim\limits_{x\to 2-}f(x)=\lim\limits_{x\to 2+}f(x)$)

II. f has jump discontinuity at $x=4$. ($\ast \lim\limits_{x\to 2-}f(x)\neq \lim\limits_{x\to 2+}f(x)$)

III. Since $\lim\limits_{x\to 2-}f(x)=\lim\limits_{x\to 2+}f(x)$, so $\lim\limits_{x\to 2}f(x)$ exists.

17. (C)

(A) Let $h(t)=\sqrt{1+t^2}$, then $f(x)=\int_0^{2x}h(t)dt$, so $f(x)=H(2x)-H(0)$.

$f'(x)=2h(2x)=2\sqrt{1+4x^2}$

(B) $g(x)=f(e^{3x})=\int_0^{2e^{3x}}\sqrt{1+t^2}\,dt$, so let $h(t)=\sqrt{1+t^2}$, then $g(x)=\int_0^{2e^{3x}}h(t)dt$.

From $g(x)=H(2e^{3x})-H(0)$, $g'(x)=6e^{3x}\cdot h(2e^{3x})=6e^{3x}\cdot\sqrt{1+4e^{6x}}$.

(C) At $x=0$, $f(0)=\int_0^0\sqrt{1+t^2}\,dt=0$, and from $f'(x)=2\sqrt{1+4x^2}$, so $f'(0)=2$.

Therefore, $y=2x$.

18. (D)

(A) The area of R is $R = \int_0^2 (x - \frac{1}{2}x^2)dx$.

(B) The volume is $\int_0^2 \cos(\frac{2\pi}{3}x)dx$.

(C) The volume is $\pi \int_0^2 \left\{ (\frac{1}{2}x^2 - 2)^2 - (x - 2)^2 \right\} dx$.

(D) The volume is $2\pi \int_0^2 (x^2 - \frac{1}{2}x^3)dx$. (Shell Method)

19. (D)

I. $\dfrac{\displaystyle\int_2^4 f'(x)dx}{4-2} = \dfrac{1}{2}(f(4)-f(2)) = \dfrac{13-5}{2} = 4$

II. $\dfrac{f(4)-f(2)}{4-2} = \dfrac{13-5}{2} = 4$.

III. By Intermediate Value Theorem, since $f(2)=5$ and $f(4)=13$, c that satisfies $f(c)=7$ exists at least 1 between 2 and 4.

20. (B)

From $f(x) = G(x) - G(a)$, $f'(x) = g(x)$.

That is, the graph given is the graph of f', from a to b, f' is negative, so f is decreasing, and from b to c, it is positive, so f is increasing.

From a to 0, f' is decreasing, so f is concave down, and from 0 to c, f' is increasing, so f is concave up.

21. (C)

(A) The object changes direction at $t=4$ and $t=8$.

(B) The total distance the object travels during the time interval $0 \le t \le 12$ is 50.

(C) $x(4) - x(0) = \int_0^4 v(t)dt$, so $\int_0^4 v(t)dt = \frac{1}{2} \times 4 \times 10 = 20$, and $x(0) = 3$,

so $x(4) = 3 + 20 = 23$.

(D) At $t=2$, it is not differentiable, so false.

22. (A)

$$\int_{-6}^{3} f(x)dx = S_1, \int_{-3}^{0} f(x)dx = -S_2, \int_{0}^{5} f(x)dx = S_3$$

Therefore $3\int_{-6}^{-3} f(x)dx - 2\int_{-3}^{0} f(x)dx + 2\int_{0}^{5} f(x)dx$

$$= 3S_1 + 2S_2 + 2S_3$$

23. (D)

From $\frac{dy}{dx} = 4x + 2$, $\int 1 dy = \int (4x+2)dx$

$\Rightarrow$ From $y = 2x^2 + 2x + C$, $4 = 18 - 6 + C$, so $C = -8$.

Therefore, $y = 2x^2 + 2x - 8$.

24. (B)

From $\int \dfrac{1}{y} dy = \int a dx$, $\ln|y| = ax + C$,

$|y| = e^{ax + C} \quad \Rightarrow \quad y = \pm e^{ax} \cdot e^{C}$

e^{C} is constant, so what can be y is $y = 3e^{ax}$.

25. (D)

Interval where f is increasing is where $f' > 0$, and interval where f is concave up is where f' is increasing.

In $f'(x) = x^2 - 4x + 3$, interval where $f'(x) > 0$ is $x > 3$ or $x < 1$.

In $f''(x) = 2x - 4 > 0$, f is concave up when $x > 2$.

Therefore, $x > 3$.

26. (C)

$\int y dy = \int \dfrac{1}{x} \ln x dx$, let $\ln x = u$, then $\dfrac{1}{x} = \dfrac{du}{dx}$.

$\dfrac{1}{2} y^2 = \int \dfrac{1}{x} u x du \quad \Rightarrow \quad \dfrac{1}{2} y^2 = \dfrac{1}{2} u^2 + C$, therefore $\dfrac{1}{2} y^2 = \dfrac{1}{2} (\ln x)^2 + C$.

$\dfrac{1}{2} = \dfrac{1}{2} + C$, so $C = 0$.

Therefore, $\dfrac{1}{2} y^2 = \dfrac{1}{2} (\ln x)^2$, since $y = \pm \ln x$, $y = \ln x$.

27. (B)

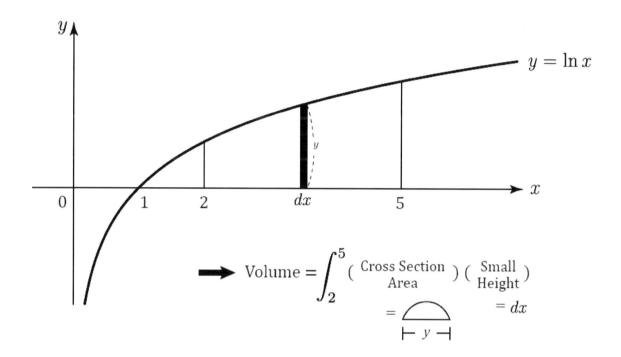

$\Rightarrow$ Volume $= \displaystyle\int_{2}^{5} \frac{1}{2}\pi(\frac{1}{2}y)^2 dx = \frac{\pi}{8}\int_{2}^{5} y^2 dx = \frac{\pi}{8}\int_{2}^{5}(\ln x)^2 dx$.

28. (A)

$3(xy+x^2)^2 \cdot (x'y+xy'+2xx') = 8y'$

$\Rightarrow 3(xy+x^2)^2(y+x\dfrac{dy}{dx}+2x) = 2\dfrac{dy}{dx}$, and substitute $x=y=1$, then

$3(2)^2(1+\dfrac{dy}{dx}+2) = 8\dfrac{dy}{dx}$, so $36+12\dfrac{dy}{dx} = 8\dfrac{dy}{dx}$.

Therefore, $\dfrac{dy}{dx}=-9$.

29. (A)

From $f(x) = G(x) - G(0)$, $f'(x) = g(x)$.

Therefore, we can infer the graph of f from the graph of $f' = g$, as follows.

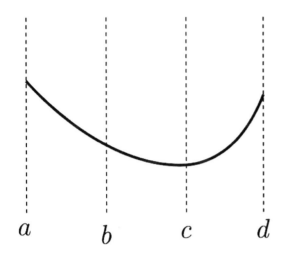

$f(d)$ or $f(a)$ can be the absolute maximum.

From $f(d) - f(a) = \int_a^d f'(x)dx < 0$, $f(d) < f(a)$.

Therefore, the answer is (A) a.

30. (B)

$\int \dfrac{1}{2\sqrt{1 - (\frac{1}{2}x)^2}} dx$, and let $\dfrac{1}{2}x = u$, then

$\dfrac{1}{2}\int \dfrac{1}{\sqrt{1 - u^2}} dx$, $\dfrac{1}{2} = \dfrac{du}{dx}$, so $\dfrac{1}{2}\int \dfrac{1}{\sqrt{1 - u^2}} 2du$

Therefore $\int \dfrac{1}{\sqrt{1 - u^2}} du = \sin^{-1} u + C$

Therefore $\sin^{-1}(\dfrac{1}{2}x) + C$.

Part B

76. (D)

From $\displaystyle\int \frac{1}{\cos^2 x} \cdot e^{\tan x} dx \implies \int \sec^2 x \cdot e^{\tan x} dx$, let $\tan x = u$, then $\sec^2 x = \dfrac{du}{dx}$.

$\displaystyle\int \sec^2 x \cdot e^u \cdot \frac{du}{\sec^2 x} \implies \int e^u du = e^u + C$

Therefore, $e^{\tan x} + C$.

77. (A)

$\displaystyle\lim_{x \to 2^-} \tan(f(x)) = \tan(2) \approx -2.185$

$\displaystyle\lim_{x \to 2^+} \tan(f(x)) = \tan(2) \approx -2.185$

78. (A)

We can draw the f' graph using calculator, as follows.

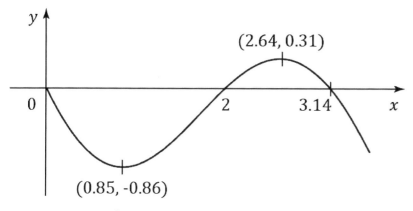

Interval where f' graph is both $f' > 0$ and increasing is $2 < x < 2.64$.

79. (A)

From $f'(x) = (2x+1) \cdot \cos(x^2+x)$, $f'(3) = 7\cos 12$,

and from $g'(x) = -x^3 - x^2 - 3$, $g'(k) = -3k^2 - 2k$

Since the two tangent lines are perpendicular, the multiple of slope is -1.

That is, $f'(3) \cdot g'(k) = -1$, therefore $-3k^2 - 2k = -\dfrac{1}{7\cos 12}$.

Therefore, $k \approx 0.056$.

80. (D)

(A) Since function f is continuous at $-1 \le x \le 3$, the minimum value exists.

(B) Function f is continuous at $-1 \le x \le 3$ and $f(-1) = -2$, $f(3) = 6$. Therefore, by Intermediate Value Theorem, c that satisfies $f(c) = 0$ at least 1 exist within the given interval.

(C) By Mean Value Theorem, $f'(c) = \dfrac{6-(-2)}{3-(-1)} = 2$.

(D) Since there are no condition like $f(-1) = f(3)$, c that satisfies $f'(c) = 0$ may exist or not exist in the given interval.

81. (A)
We can draw the graph using calculator as follows.

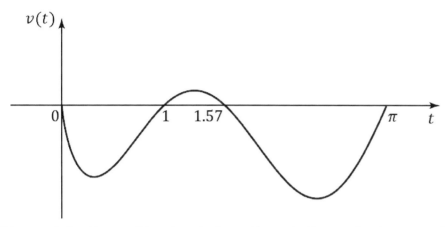

We can infer the position graph from the velocity graph above, as follows.

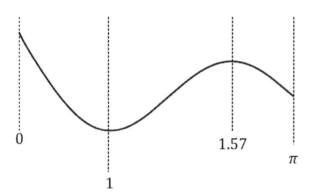

Graph of Position

Therefore, candidates of the absolute maximum of position is $t = 0$ and $t = 1.57$.

$X(1.57) - X(0) = \int_0^{1.57} v(t) dt = -0.37 < 0$, therefore $X(1.57) < X(0)$.

Therefore, the particle is farthest to the right at $t = 0$.

82. (A)

$$\frac{d^3y}{dx^3} = \boxed{\frac{d}{dx}}\,\boxed{\frac{d}{dx}}\,\boxed{\frac{dy}{dx}}$$

③ ② ①

① $\dfrac{dy}{dx} = 3\cos 3x + 5e^{5x}$

② $\dfrac{d}{dx}\dfrac{dy}{dx} = \dfrac{d}{dx}(3\cos 3x + 5e^{5x}) = -9\sin 3x + 25e^{5x}$

③ $\dfrac{d}{dx}\dfrac{d}{dx}\dfrac{dy}{dx} = \dfrac{d^3y}{dx^3} = \dfrac{d}{dx}(-9\sin 3x + 25e^{5x}) = -27\cos 3x + 125e^{5x}$

83. (D)

(A) Let the acceleration as $a(t)$, then at time $t = 3$, $a(3) \approx 1$.

(B) Average Speed $= \dfrac{1}{8}\displaystyle\int_0^8 |v(t)|\,dt \approx 1.74$

(C) From $x(5) - x(0) = \displaystyle\int_0^5 v(t)\,dt$, $x(5) = 10 + \displaystyle\int_0^5 v(t)\,dt \approx 14.84$

(D) Velocity is positive at $0 < t < 1$, and acceleration is negative at $0 < t < 1$.
Therefore, in the given interval, speed is decreasing.

84. (B)

From $g(x) = F(2x) - F(-4)$, $g'(x) = 2f(2x)$.

(A) $g'(2) = 2f(4) = 2$

(B) From $g''(x) = 4f'(2x)$, $g''(1.5) = 4f'(3)$.

$f'(3) = \dfrac{1 - (-3)}{4 - 2} = \dfrac{4}{2} = 2$

Therefore, $g''(1.5) = 4 \cdot 2 = 8$

(C) $g(0) = \displaystyle\int_{-4}^0 f(t)\,dt = \dfrac{1}{2} \cdot 2 \cdot 3 + \dfrac{1}{2}(3+5) \cdot 2 = 11$

(D) Since $g'(1) = 2f(2)$ and f changes from decreasing to increasing at $x = 2$, g has inflection point at $x = 1$.

85. (A)

From $g(x) = F(x) - F(1)$, $g'(x) = f(x)$.

When $g' = f < 0$ and $g' = f$ is increasing, g is both decreasing and concave up.

86. (C)

Let $\sin\left(\dfrac{3}{2t^2}\right) = g(t)$, then from $f(x) = G(x) - G(1)$, $f'(x) = g(x) = \sin\left(\dfrac{3}{2x^2}\right)$.

Using calculator, let's draw the graph of $f'(x)$.

Then, sign of f' changes from positive to negative at $x = 1.45$ and $x = 2.51$.

Therefore, function f has local maximum at $x = 1.45$ and $x = 2.51$.

87. (B)

By Intermediate Value Theorem,

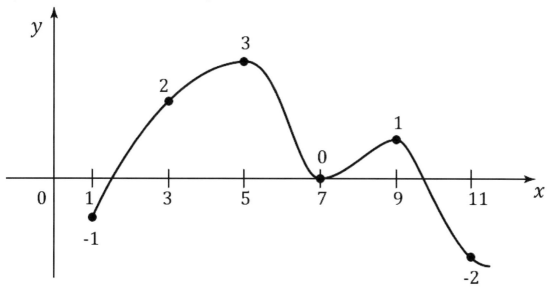

This meets $x-$axis at least 3 times.

88. (A)

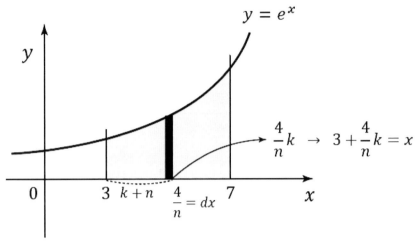

From the figure above,

① $\dfrac{4}{n} - dx$ ② $3 + \dfrac{4}{n}k = x$, therefore $\displaystyle\lim_{n \to \infty} \sum_{k=1}^{n} e^{(3 + \frac{4}{n}k)} \cdot \dfrac{4}{n}$

89. (C)

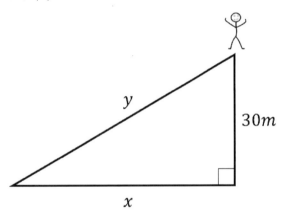

From $y^2 = x^2 + 30^2$, since $2y\dfrac{dy}{dt} = 2x\dfrac{dx}{dt}$, $\dfrac{dy}{dt} = \dfrac{x}{y}\dfrac{dx}{dt}$.

$t = 5$ seconds, therefore $x = 5 \times 10 = 50$.

Then, by $y^2 = 30^2 + 50^2$, $y \approx 58.31$ and $\dfrac{dx}{dt} = 10$.

Therefore, $\dfrac{dy}{dt} = \dfrac{50}{58.31} \times 10 \approx 8.57$.

90. (B)

$g'(5) = (f^{-1})'(5) = \dfrac{1}{f'(1)} = \dfrac{1}{2}$

AP Calculus AB
💡 Practice Test 3

CALCULUS AB
SECTION I, Part A
Time-60 minutes
Number of questions-30

A CALCULATOR MAY NOT BE USED ON THIS PART OF THE EXAMINATION.

Directions: Solve each of the following problems, using the available space for scratchwork. After examining the form of the choices, decide which is the best of the choices given and fill in the corresponding oval on the answer sheet. No credit will be given for anything written in the test book. Do not spend too much time on any one problem.

In this test:

(1) Unless otherwise specified, the domain of a function is assumed to be the set of all real numbers x for which $f(x)$ is a real number.

(B) The inverse of a trigonometric function f may be indicated using the inverse function notation f^{-1} or with the prefix "arc" (e.g., $\sin^{-1} = \arcsin x$).

1. $\displaystyle\lim_{x \to \infty} \frac{3^{x-1}+1}{3^{x+1}+1}$ is

(A) $\dfrac{1}{9}$

(B) -2

(C) 2

(D) 3

2. $\displaystyle\int \frac{\ln x}{x}dx =$

(A) $\ln x^2 + C$

(B) $\dfrac{1}{2}(\ln x)^2 + C$

(C) $x^{-1} + C$

(D) $-x^{-1} + C$

3. If $\dfrac{dy}{dx} = \sqrt{3+x^2}$ then $\dfrac{d^2y}{dx^2} =$

(A) x

(B) $\sqrt{3+x^2}$

(C) $\dfrac{1}{\sqrt{3+x^2}}$

(D) $\dfrac{x}{\sqrt{3+x^2}}$

4. $\displaystyle\int \sin(3x)\,dx =$

(A) $-3\cos(3x) + C$

(B) $-\dfrac{1}{3}\cos x + C$

(C) $-3\cos x + C$

(D) $-\dfrac{1}{3}\cos(3x) + C$

5. $\lim\limits_{x \to 1} \dfrac{|x-1|}{x^2 - 3x + 2}$ is

(A) Nonexistent

(B) -1

(C) 0

(D) 1

6. $f(x) = \begin{cases} \dfrac{x^2 + 3x + 2}{x+1} & (x \neq -1) \\ 2 & (x = -1) \end{cases}$

Let f be the function defined above. Which of the following statements about f are true?

I. f has a limit at $x = -1$

II. f is continuous at $x = -1$

III. f is differentiable at $x = -1$

(A) I only

(B) II only

(C) III only

(D) I and II only

7. A particle moves along the y-axis with acceleration of $a(t)=2$ and when time $t=1$, the velocity is -2. The particle is at position $y=1$ when time $t=0$. what is the position of the particle at time $t=2$?

(A) -3

(B) -1

(C) 0

(D) 2

8. $\displaystyle\int \frac{x+1}{x^2+2x+5}\, dx =$

(A) x^2+2x+5

(B) $e^{x^2+2x+5}+C$

(C) $2\ln|x^2+2x+5|+C$

(D) $\dfrac{1}{2}\ln|x^2+2x+5|+C$

9. If $f(x) = e^{\tan(2x)}$, then $f'(x) =$

(A) $2e^{(2/x)} \ln x$

(B) $2e^{\tan 2x} \sec^2 2x$

(C) $e^{(-2/x^2)}$

(D) $-\dfrac{1}{x^2} e^{(2/x)}$

10. The function of f is given by $f(x) = \sqrt[3]{x-2}$. Which value of x makes f attain its critical value of x?

(A) 2

(B) 3

(C) 4

(D) 5

x	1.8	2.2	2.4	2.6
$f(x)$	5	7	11	20

11. The function of f is differentiable and has values as shown in the table above which of the following could be the value of $f'(2)$?

(A) 5

(B) 8

(C) 12

(D) 18

12. If $\tan(xy) = 2x$, then $\dfrac{dy}{dx} =$

(A) $\dfrac{1}{\cos(xy)}$

(B) $\dfrac{1}{x\cos(xy)}$

(C) $\dfrac{1-\cos(xy)}{\cos(xy)}$

(D) $\dfrac{2\cos^2(xy)-y}{x}$

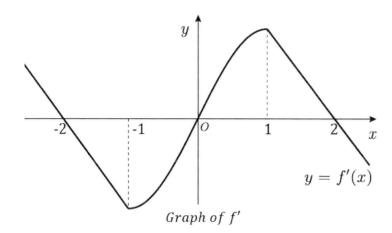

Graph of f'

13. The graph of f', the derivative of the function f, is shown above. Which of the following statements about f are true?

I. f is both concave up and increasing for $0 < x < 1$.

II. f is not differentiable at $x = -1$ and 1.

III. f has a local minimum at $x = 0$.

(A) I only

(B) III only

(C) I and II only

(D) I and III only

14. A water jar has a height of 15cm. The area, in square centimeters, of the horizontal cross section at height h centimeter is modeled by the function w given by $w(h) = e^{2h} + h$. Based on this model, what is the volume of the water jar?

(A) $e^{30} + 225$

(B) $\dfrac{1}{2}e^{30}$

(C) $\dfrac{1}{2}e^{30} + 112$

(D) $\dfrac{1}{2}e^{30} + \dfrac{225}{2}$

15. If $f(x) = x^2$ and $g(x) = \sin(2x)$, then $(f \circ g)'$ is

(A) $2\sin(4x)$

(B) $2\sin(2x)$

(C) $2\cos(2x)$

(D) $2\sin(2x)\sin(4x)$

x	1	2	3	4	5
$f'(x)$	2	-3	0	2	-2

16. The polynomial function f has selected values of its first derivative f' given in the table above. Which of the following statements must be true?

(A) f has relative maximum at $x = 4$

(B) f has a point of inflection at $x = 3$

(C) f is increasing on the interval $(3,4)$

(D) f has at least one local maximum on the interval $(1,2)$

17. What is(are) horizontal asymptotes of the graph of $y = \dfrac{2 + 3^{x+1}}{2 - 3^x}$ in the xy-plane?

(A) $y = 1$

(B) $y = -3$

(C) $y = 0$ and $y = 1$

(D) $y = -3$ and $y = 1$

18. $\dfrac{d}{dx} \displaystyle\int_{1}^{x^2} \sec t\, dt =$

(A) $2x \sec x$

(B) $2x \sec x^2$

(C) $2x^2 \sec x^2$

(D) $\sec x$

19. A curve has slope of $x - 2$ at each point (x, y) on the curve. Which of the following is an equation for this curve if it passes through the point $(2, -2)$?

(A) $y = x^2 - x$

(B) $y = x^2 - 2x$

(C) $y = \dfrac{1}{2}x^2 - 2x + 1$

(D) $y = \dfrac{1}{2}x^2 - 2x$

20. Let f be a function with second derivative given by $f''(x) = x^2(x-1)^3(x-4)^4$. What is the x-coordinate of the points of inflection of the graph of f?

(A) 0 only

(B) 1 only

(C) 1 and 4 only

(D) 0 and 4 only

21. A student wants to design an open box having a square base and a surface area of 24 square inches. What dimensions will produce a box with maximum volume?

(A) $2\sqrt{2} \times 2\sqrt{2} \times \sqrt{2}$

(B) $\sqrt{2} \times \sqrt{2} \times \dfrac{11\sqrt{2}}{4}$

(C) $2 \times 2 \times \dfrac{5}{2}$

(D) $1 \times 1 \times \dfrac{23}{4}$

22. If $y = xy + y^2 + 5$, then $y = 1$, $\dfrac{dy}{dx}$ is

(A) $\dfrac{1}{4}$

(B) $\dfrac{1}{2}$

(C) 1

(D) 2

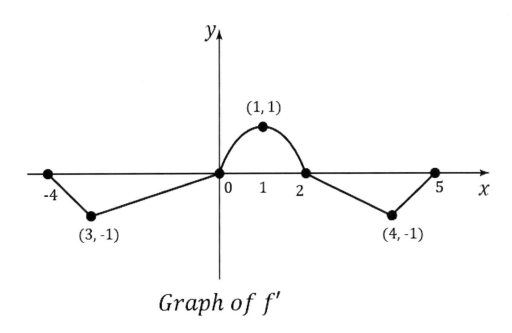

Graph of f′

23. The function f is differentiable on the closed interval $[-4, 5]$ and satisfies $f(4)=3$. The graph of f', the derivative of f, consists of a semicircle and four line segments, as shown in the figure above. What is the absolute maximum value of f on the closed interval $[-4, 5]$?

(A) $6 - \dfrac{\pi}{2}$

(B) $\dfrac{\pi}{2}$

(C) $-3 + \dfrac{\pi}{2}$

(D) 6

24. $f(x) = \begin{cases} ax^2 + b & \text{for } x \leq 1 \\ x^3 + 3ax & \text{for } x > 1 \end{cases}$

Let f be the function defined above, where a and b are constants. If f is differentiable at $x = 1$, what is the value of $a + b$?

(A) -8

(B) -4

(C) 0

(D) 2

25. Let f be a function that is continuous for all real numbers. The table below gives values of f for selected point in the closed interval $[1, 12]$.

x	1	3	5	6	8	12
$f(x)$	2	8	2	-4	1	7

Use a right Riemann sum with sub-intervals indicated by the data in the table to approximate $\int_1^{12} f(x)\,dx$.

(A) 24

(B) 28

(C) 32

(D) 46

x	$f(x)$	$f'(x)$	$g(x)$	$g'(x)$
-2	2	4	3	5
1	1	-2	-1	7
5	3	5	-2	2

26. The table above gives values of f, and f', and g and g' at selected of x.

If $h(x) = (f \circ g)(x)$, the $h'(5) =$

(A) 4

(B) 8

(C) 16

(D) 24

27. Let f be a function such that $\lim\limits_{h \to 0} \dfrac{f(h) - f(0)}{h} = 3$. Which of the following could be false?

(A) $\lim\limits_{x \to 0} f(x)$ exists

(B) $f(0)$ exists

(C) f' is continuous at $x = 0$

(D) $\lim\limits_{x \to 0} \dfrac{f(x) - f(0)}{x} = 3$

28. The acceleration of a car driving in a straight road is given in terms of time t by $a(t)=2$. If the velocity of the car is 2 at $t=1$ and if $P(t)$ is the distance of the car from the origin at time t, what is $P(3)-P(1)$?

(A) 0

(B) 2

(C) 4

(D) 8

29. The function f is continuous on the closed interval $[1,3]$ and differentiable on the open interval $(1,3)$. If $f'(2)=-1$ and $f''(x)>0$ on the open interval $(1,3)$, which of the following could be a table of values for f? ($\ast f$ has no inflection points)

(A)

x	$f(x)$
1	3
2	2
3	1.5

(B)

x	$f(x)$
1	3.5
2	2
3	1.5

(C)

x	$f(x)$
1	4
2	2
3	1

(D)

x	$f(x)$
1	2
2	2.5
3	3.5

223

x	$f(x)$	$f'(x)$
1	0	1
2	1	2
3	3	5

30. The table above gives selected values for a differentiable and function f and its derivative. If g is the inverse function of f, what is the value of $g'(1)$?

(A) -5

(B) -1

(C) 0

(D) $\dfrac{1}{2}$

A GRAPHING CALCULATOR IS REQUIRED FOR SOME QUESTIONS ON THIS PART OF THE EXAMINATION.

Directions: Solve each of the following problems, using the available space for scratchwork. After examining the form of the choices, decide which is the best of choices given and fill in the corresponding oval on the answer sheet. No credit will be given for anything written in the test book. Do no spend too much time on any one problem.

BE SURE YOU ARE USING PAGE 3 OF THE ANSWER SHEET TO RECORD YOUR ANSWERS TO QUESTIONS NUMBERED 76-90.

YOU MAY NOT RETURN TO PAGE 2 OF THE ANSWER SHEET.

In this test:

(1) The exact numerical value of the correct does not always appear among the choices given. When this happens, select from among the choices the number that best approximates the exact numerical value.

(2) Unless otherwise specified, the domain of a function f is assumed to be the set of all real numbers x for which $f(x)$ is real number.

(3) The inverse of a trigonometric function f may be indicated using the inverse function notation f^{-1} or with the prefix "arc" (e.g, $\sin^{-1}x = \arcsin x$)

76. What is the average value of $y = \sin(x^3)$ on the closed interval $[1, 3]$?

(A) 0.532

(B) 0.111

(C) 0.223

(D) 0.364

77. If $\lim\limits_{x \to 2} f(x) = 3$, which of the following must be true?

I. $\lim\limits_{x \to 2^-} f(x) = \lim\limits_{x \to 2^+} f(x)$

II. f is continuous at $x = 2$

III. $f(2) = 3$

(A) I only

(B) I and II only

(C) I and III only

(D) II and III only

78. Evaluate $\displaystyle\lim_{n\to\infty}\sum_{k=1}^{n}(1+\frac{2k}{n})^2 \cdot \frac{1}{n}$

(A) 2

(B) 4

(C) $\dfrac{27}{6}$

(D) $\dfrac{13}{3}$

79. If $\displaystyle\int_{1}^{7}f(x)dx = 8$ and $\displaystyle\int_{7}^{2}f(x)dx = 5$, then $\displaystyle\int_{1}^{2}f(x)dx =$

(A) -3

(B) 0

(C) 3

(D) 13

80. Andy who is 1.8 meters tall walks directly away from a streetlight that is 5.4 meters above the ground. If Andy is walking at a constant rate and his shadow is lengthening at the rate of 0.5 meters per second, at what rate, in meters per second, is Andy walking?

(A) 1.0

(B) 1.2

(C) 1.6

(D) 2.4

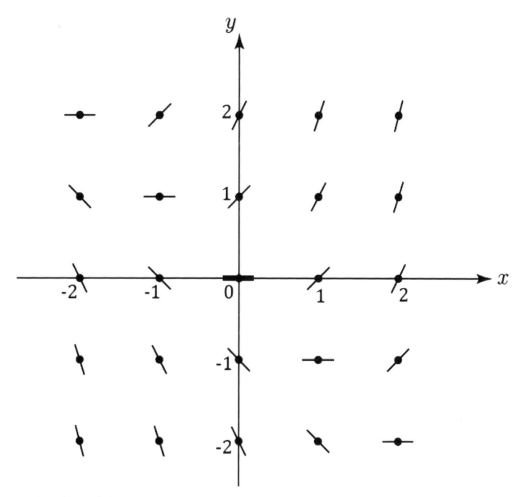

81. The slope field shown above corresponds to which of the following differential equation?

(A) $\dfrac{dy}{dx} = x + y$

(B) $\dfrac{dy}{dx} = x - y$

(C) $\dfrac{dy}{dx} = x^2 + y$

(D) $\dfrac{dy}{dx} = x + y^2$

82. What is the slope of the line tangent to the graph of $\ln(x^2y^3) = 2y$ at the point $(e,\ 1)$?

(A) $-\dfrac{2}{e}$

(B) $-\dfrac{1}{2}$

(C) 0

(D) $\dfrac{1}{e}$

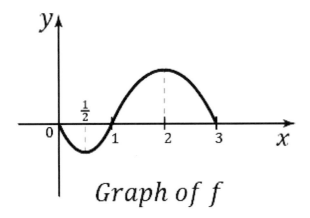

Graph of f

83. The graph of the differentiable function f is shown in the figure above. Let g be the function defined by $g(x) = \displaystyle\int_0^x f(t)\,dt$. Which of the following is true?

(A) g is decreasing at $x = 2$.
(B) f has an inflection points at $x = 1$ and $x = 2$.
(C) f has a local minimum at $x = 1$.

(D) g has an inflection points at $x = \dfrac{1}{2}$ and $x = 2$.

84. Let f be the function given by $f(x) = \int_0^{x^2} \cos(t)dt$ for $1 \le x \le 5$. How many points of relative minimum does the graph of f have on the interval $[1, 5]$?

(A) Three

(B) Four

(C) Five

(D) Six

85. $\int \dfrac{4x^2}{1+4x^2}dx =$

(A) $\ln(1+4x^2) + C$

(B) $x - \dfrac{1}{2}\tan^{-1}(2x) + C$

(C) $1 + \ln(4x^2) + C$

(D) $\dfrac{4}{3}x^3 \cdot \ln(1+4x^2) + C$

231

86. Which of the following statements must be true?

I. If function f is continuous at $x=1$, then $\displaystyle\lim_{x\to 1-}\frac{f(x)-f(1)}{x-1}=\lim_{x\to 1+}\frac{f(x)-f(1)}{x-1}$.

II. If function f is differentiable at $x=1$, then $\displaystyle\lim_{x\to 1-}f(x)=\lim_{x\to 1+}f(x)$.

III. If function f is defined at $x=1$, then $\displaystyle\lim_{x\to 1}f(x)=f(1)$.

(A) I only
(B) II only
(C) I and II only
(D) II and III only

87. A solid with a circular base with equation $x^2+y^2=25$, and all cross sections parallel to the $x-$axis are isosceles right triangle with hypotenuse lying along the base. What is the volume of this solid?

(A) $\dfrac{125}{3}$

(B) 50

(C) $\dfrac{250}{3}$

(D) $\dfrac{500}{3}$

88. Truck is traveling west at $70\mathrm{km/hour}$ and sedan is traveling north at $90\mathrm{km/hour}$. Both are headed for the intersection of the two roads. At what rate are the cars approaching each other when Truck is $50m$ and Sedan is $100m$ from the intersection?

(A) $-126.4\mathrm{km/h}$

(B) $-111.8\mathrm{km/h}$

(C) $-91.6\mathrm{km/h}$

(D) $-32.7\mathrm{km/h}$

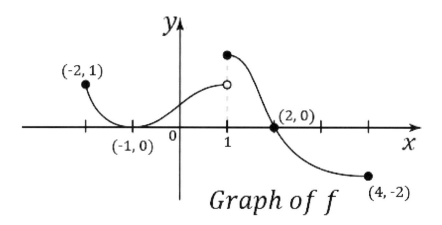

Graph of f

89. The function f is defined on the closed inteval $[-2, 4]$. The graph of f is shown in the figure above. If g be the function defined by $g(x) = \int_{-1}^{x} f(t)dt$, which of the following must be true?

(A) f has a removable discontinuity at $x = 1$.

(B) g is differentiable at $x = 1$.

(C) g has a local minimum at $x = -1$.

(D) f has an inflection point at $x = -1$.

90. The wild rabbit population P in a forest grows according to the equation $\dfrac{dP}{dt} = kP$, where k is a constant and t is measured in years. If the wild rabbit population triples every 10 years, what is the value of k?

(A) 0.11

(B) 0.23

(C) 0.36

(D) 0.41

AP Calculus AB Practice Test 3
❯ Answer Key

Part A

1. (A)	6. (A)	11. (A)	16. (D)	21. (A)	26. (B)
2. (B)	7. (A)	12. (D)	17. (D)	22. (A)	27. (C)
3. (D)	8. (D)	13. (D)	18. (B)	23. (A)	28. (D)
4. (D)	9. (B)	14. (C)	19. (D)	24. (A)	29. (B)
5. (A)	10. (A)	15. (A)	20. (B)	25. (D)	30. (D)

Part B

76. (B)	79. (D)	82. (A)	85. (B)	88. (B)
77. (A)	80. (A)	83. (D)	86. (B)	89. (B)
78. (D)	81. (A)	84. (B)	87. (D)	90. (A)

PART A

1. (A)

$$\lim_{x\to\infty}\frac{\frac{1}{3}\times 3^x+1}{3\times 3^x+1}=\lim_{x\to\infty}\frac{\frac{1}{3}+\frac{1}{3^x}}{3+\frac{1}{3^x}}=\lim_{x\to\infty}\frac{\frac{1}{3}+\frac{1}{\infty}}{3+\frac{1}{\infty}}=\frac{\frac{1}{3}}{3}=\frac{1}{9}$$

2. (B)

$$\int\frac{1}{x}\ln x\,dx\Rightarrow \ln x=u,\frac{1}{x}=\frac{du}{dx}\Rightarrow\int\frac{1}{x}ux\,du=\int u\,du$$

$$\Rightarrow\frac{1}{2}u^2+C\Rightarrow\frac{1}{2}(\ln x)^2+C$$

3. (D)

$$\frac{d^2y}{dx^2}=\frac{d}{dx}\frac{dy}{dx}=\frac{d}{dx}(\sqrt{3+x^2})=\frac{d}{dx}(3+x^2)^{\frac{1}{2}}=\frac{1}{2}(3+x^2)^{\frac{1}{2}-1}\cdot 2x$$

$$\Rightarrow \frac{x}{\sqrt{3+x^2}}$$

4. (D)

$$3x=u\Rightarrow 3=\frac{du}{dx}$$

$$\Rightarrow\frac{1}{3}\int\sin u\,du=-\frac{1}{3}\cos u+C=-\frac{1}{3}\cos(3x)+C$$

5. (A)

$$\lim_{x \to 1-} \frac{-(x-1)}{(x-1)(x-2)} = \lim_{x \to 1-} \frac{-1}{x-2} = 1$$

$$\lim_{x \to 1+} \frac{(x-1)}{(x-1)(x-2)} = \lim_{x \to 1+} \frac{1}{x-2} = -1$$

$$\lim_{x \to 1-} \frac{|x-1|}{x^2 - 3x + 2} \neq \lim_{x \to 1+} \frac{|x-1|}{x^2 - 3x + 2}, \text{ therefore } \lim_{x \to 1} \frac{|x-1|}{x^2 - 3x + 2} \text{ is nonexisent.}$$

6. (A)

I. $$\lim_{x \to -1-0} \frac{x^2 + 3x + 2}{x+1} = \lim_{x \to -1+0} \frac{x^2 + 3x + 2}{x+1} = 1$$

II. $$\lim_{x \to -1} \frac{x^2 + 3x + 2}{x+1} = 1 \neq f(-1) = 2$$

III. Function f is discontinuous at $x = -1 \Rightarrow$ not differentiable at $x = -1$

7. (A)

Velocity = $\int 2dt = 2t + C$, $C = -4$. $(\because V(1) = -2)$

Velocity $v(t) = 2t - 4$

Position $P(2) = P(0) + \int_0^2 (2t - 4)dt = -3$

8. (D)

$$x^2 + 2x + 5 = u \Rightarrow 2(x+1) = \frac{du}{dx}$$

$$\Rightarrow \int \frac{x+1}{u} \times \frac{1}{2(x+1)} du = \frac{1}{2} \int \frac{1}{u} du$$

$$= \frac{1}{2} \ln|x^2 + 2x + 5| + C$$

9. (B)

$f'(x) = e^{\tan(2x)}(\tan 2x)' \Rightarrow 2\sec^2(2x)e^{\tan(2x)}$

10. (A)

Critical Value

(1) $x - values$ that $f'(x) = 0$

(2) $x - values$ that $f'(x)$ is undefined

$f'(x) = \dfrac{1}{3\sqrt[3]{(x-2)^2}} \Rightarrow f'(2)$ is undefined.

Therefore, $x = 2$

11. (A)

For the problem that asks for $f'(x)$ when the table and the condition that it is differentiable is given, use Mean Value Theorem!

$f'(2) = \dfrac{7-5}{2.2-1.8} = 5$

12. (D)

$\sec^2(xy)(x'y + xy') = 2 \Rightarrow y' = \dfrac{dy}{dx} = \dfrac{2}{x\sec^2(xy)} - \dfrac{y}{x} = \dfrac{2\cos^2(xy) - y}{x}$

13. (D)

I. In the interval where f' is increasing, f is concave up, and in the interval where f' is positive, f is increasing. Therefore, True.

II. Since $f'(-1)$ and $f'(1)$ exist, f is differentiable at $x=-1$ and $x=1$. Therefore, false.

III. Local minimum is where f' changes from negative to positive. Therefore, true.

14. (C)

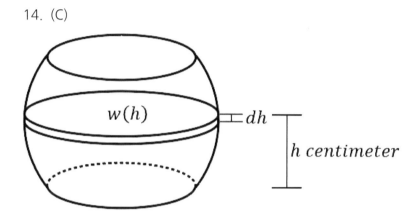

In the figure above,

$$V = \int_0^{15} (e^{2h} + h)\,dh = [\frac{1}{2}e^{2h} + \frac{1}{2}h^2]_0^{15}$$

$$= \frac{1}{2}e^{30} + \frac{225}{2} - \frac{1}{2}$$

$$= \frac{1}{2}e^{30} + \frac{224}{2} = \frac{1}{2}e^{30} + 112$$

15. (A)

$f'(x) = 2x, \ g'(x) = 2\cos(2x)$

$(f \circ g)' = f'(g(x))g'(x) = 2 \times 2\sin(2x)\cos(2x) = 2\sin(4x)$ (※ $\sin 2x = 2\sin x \cos x$)

16. (D)
Graph like below, can be drawn.

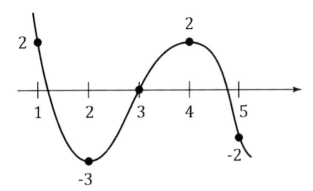

Graph of f'

Function f is a polynomial function, therefore it is always continuous.

Also, f' satisfies $f'(1) > 0$ and $f'(2) < 0$, therefore within the interval of $(1, 2)$, at least at one point, does f' changes from positive to negative.

Therefore, at least 1 local maximum is in interval $(1, 2)$.

(C) Graph like below, can be drawn.

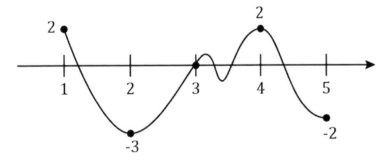

Therefore, f can be decreasing on the interval $(3, 4)$. Thus, (C) is false.

17. (D)
Horizontal Asymptote

① $\lim\limits_{x \to +\infty} \dfrac{2+3^{x+1}}{2-3^x} = \lim\limits_{x \to \infty} \dfrac{2+3 \cdot 3^x}{2-3^x} = \lim\limits_{x \to \infty} \dfrac{\dfrac{2}{3^x}+3}{\dfrac{2}{3^x}-1} = -3$

② $\lim\limits_{x \to -\infty} \dfrac{2+3^{x+1}}{2-3^x} = \dfrac{2+3^{-\infty}}{2-3^{-\infty}} = 1$

Therefore, horizontal asymptote is $y = -3$ and $y = 1$.

18. (B)

$f(t) = \sec t, \dfrac{d}{dx}\displaystyle\int_1^{x^2} f(t)dt \Rightarrow \dfrac{d}{dx}(F(x^2) - F(1))$

$\Rightarrow 2xf(x^2) = 2x\sec x^2$

19. (D)

$\dfrac{dy}{dx} = x - 2 \Rightarrow \displaystyle\int dy = \int (x-2)dx \Rightarrow y = \dfrac{1}{2}x^2 - 2x + C, \; C = 0$

$\therefore y = \dfrac{1}{2}x^2 - 2x$

20. (B)

$f''(x) = x^2(x-1)^3(x-4)^4$ is drawn as below.

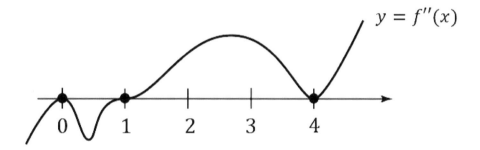

Graph of f''

Like the figure above, sign change of f'' is at $x = 1$.
Therefore, function f has inflection point at $x = 1$.

21. (A)

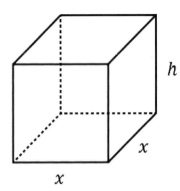

In the figure above, Volume is $x^2 h$, and surface area satisfies $x^2 + 4xh = 24$.

Since $h = \dfrac{24 - x^2}{4x}$,

Volume $= x^2 \cdot \dfrac{24 - x^2}{4x} = \dfrac{24x - x^3}{4} = 6x - \dfrac{1}{4}x^3$

Since $x > 0$ and $h > 0$, $h = \dfrac{24 - x^2}{4x} > 0 \Rightarrow 0 < x < 2\sqrt{6}$.

Let the volume as $V(x)$, then the graph of $V(x)$ is as below.

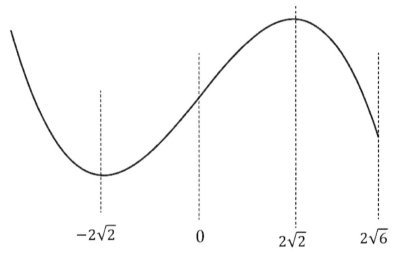

$$-2\sqrt{2} \qquad 0 \qquad 2\sqrt{2} \qquad 2\sqrt{6}$$

Graph of $V(x)$

From $V'(x) = 6 - \dfrac{3}{4}x^2 = 0$, $x = \pm 2\sqrt{2}$

Therefore, at $x = 2\sqrt{2}$, the volume would be maximum.

$h = \dfrac{24 - 8}{4 \cdot 2\sqrt{2}} = \dfrac{16}{8\sqrt{2}} = \dfrac{2}{\sqrt{2}} = \sqrt{2}$.

Therefore, when $2\sqrt{2} \times 2\sqrt{2} \times \sqrt{2}$, the box would have the maximum volume.

22. (A)

When $y = 1$, $x = -5$.

$y' = x'y + xy' + 2yy'$

$\Rightarrow \dfrac{dy}{dx} = 1 - 5\dfrac{dy}{dx} + 2\dfrac{dy}{dx}$ $\Rightarrow \dfrac{dy}{dx} = \dfrac{1}{4}$

23. (A)

We can infer f graph from f' graph, as follows.

Graph of f'

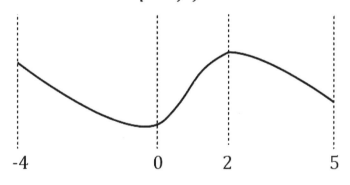

Graph of f

As in the figure above, candidates of absolute maximum are $f(-4)$ and $f(2)$.

Therefore, $f(2) - f(-4) = \displaystyle\int_{-4}^{2} f'(x)dx = \dfrac{1}{2} \cdot 4(-1) + \dfrac{1}{2}\pi = -2 + \dfrac{\pi}{2} < 0$

$\Rightarrow f(2) < f(-4)$

Therefore, absolute maximum is $f(-4)$.

Using $f(4) = 3$ which is given in the problem, we can obtain $f(-4)$.

$f(4) - f(-4) = \displaystyle\int_{-4}^{4} f'(x)dx$

$\Rightarrow f(-4) = 3 - \displaystyle\int_{-4}^{4} f'(x)dx = 3 - (-2 + \dfrac{\pi}{2} - 1) = 6 - \dfrac{\pi}{2}$

24. (A)

(1) At $x=1$, $a+b=1+3a \implies -2a+b=1$

(2) At $x=1$, $(ax^2+b)'=(x^3+3ax)' \implies 2a=3+3a \implies a=-3$

$\therefore a+b=-8$

25. (D)

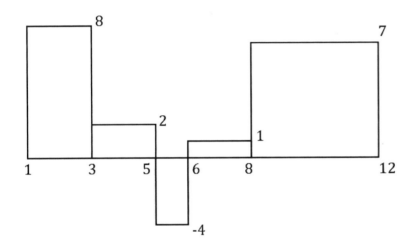

$R=(8\times2)+(2\times2)+(1\times(-4))+(2\times1)+(4\times7)=46$

26. (B)

$h'(x)=f'(g(x))g'(x) \implies h'(5)=f'(g(5))g'(5)=f'(-2)\times g'(5)=4\times2=8$

27. (C)

$$\lim_{h \to 0} \frac{f(0+h) - f(0)}{h} = f'(0) = 3$$

$\Rightarrow f(x)$ is differential at $x = 0$

$\Rightarrow f(x)$ is continuous at $x = 0$

$\Rightarrow \lim_{x \to 0} f(x)$ exist

$\Rightarrow f(0)$ exists

$$\lim_{x \to 0} \frac{f(x) - f(0)}{x - 0} = f'(0) = 3$$

f is differentiable at $x = 0$, so case as follows can be happen.

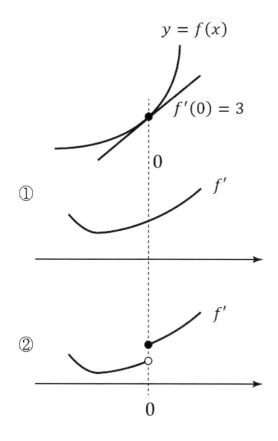

That is, the value of $f'(0)$ only needs to exist, so as like the figure ②, $f'(0)$ can be discontinuous even though it exists, at $x = 0$.

28. (D)

Velocity$= \int 2dt = 2t + C$, $v(1) = 2$, $C = 0$ ∴ $v(t) = 2t$

$$= \begin{cases} f(3) - f(1) = \int_1^3 f'(t)dt = \int_1^3 2t\,dt = \left[t^2\right]_1^3 = 8 \\ P(3) - P(1) = \int_1^3 v(t)dt = \int_1^3 2t\,dt = \left[t^2\right]_1^3 = 8 \end{cases}$$

29. (B)

The graph that satisfies the equation of $f'(2) = -1, f''(2) > 0$ will be the graph that f is Decreasing as well as Concave Up. Let's look at the pictures shown below.

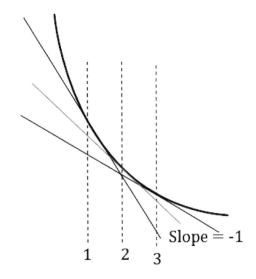

Slope $= -1$

1   2   3

Therefore the slope between 1 and 2 has to be lower than -1 and the slope between 2 and 3 has to be bigger than -1. Therefore the possible answer from below will be (B).

30. (D)

$$g'(1) = (f^{-1})'(1) = \frac{1}{f'(2)} = \frac{1}{2}$$

PART B

76. (B)

Average Value $= \dfrac{\displaystyle\int_{1}^{3} \sin(x^{3})\,dx}{3-1} \approx 0.111$

77. (A)

$\lim\limits_{x\to 2} f(x) = 3$ means that $\lim\limits_{x\to 2-} f(x) = \lim\limits_{x\to 2+} f(x)$.

Case as follows can be made :

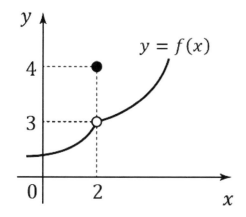

$\Rightarrow$
- $\lim\limits_{x\to 2-} f(x) = \lim\limits_{x\to 2+} f(x) = 3$
- f is discontinuous at $x = 2$.
- $f(2) = 4$

78. (D)

Draw the $y = f(x)$ graph roughly, and the scope would be from any number from the origin, but here, we set the range to 2.

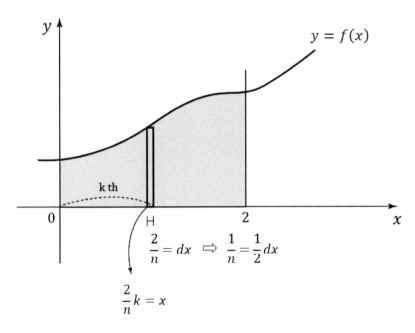

$$\boxed{\lim_{n \to \infty} \sum_{k=1}^{n} \left(1 + \frac{2k}{n}\right)^2 \cdot \frac{1}{n}}$$

$$\frac{1}{2} \int_0^2 (1+x)^2 dx \implies 1+x = u, \; 1 = \frac{du}{dx}$$

$$\implies \frac{1}{2} \int_1^3 u^2 du = \frac{1}{2}\left[\frac{1}{3}u^3\right]_1^3 = \frac{1}{2}\left(9 - \frac{1}{3}\right) = \frac{1}{2} \cdot \frac{26}{3} = \frac{13}{3}$$

79. (D)

$$\int_1^7 f(x)dx + \int_7^2 f(x)dx = \int_1^2 f(x)dx = 13$$

80. (A)

Streetlight

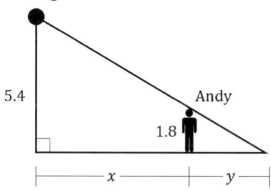

- Topic we should obtain from the figure above is $\dfrac{dx}{dt}$

- From $1.8 : 5.4 = y : x+y$, $5.4y = 1.8x + 1.8y$
$\Rightarrow$ since $3.6y = 1.8x$, $2y = x$.
$\Rightarrow$ From $2\dfrac{dy}{dt} = \dfrac{dx}{dt}$, $\dfrac{dy}{dt} = 0.5$. We obtain $\dfrac{dx}{dt} = 1$.

81. (A)
Among choices given, substitute $(2, -2)$ and $(1, -1)$ and find that has 0 as a slope, then the answer is (A).

82. (A)
$$\dfrac{1}{x^2y^3}(x^2y^3)' = (2y)' \Rightarrow \dfrac{1}{x^2y^3}(2xy^3 + 3x^2y^2\dfrac{dy}{dx}) = 2\dfrac{dy}{dx}$$

$$\Rightarrow x = e,\ y = 1 \Rightarrow \dfrac{dy}{dx} = -\dfrac{2}{e}$$

83. (D)

$g(x) = F(x) - F(0)$, so $g'(x) = f(x)$.

(A) Since $g'(2) > 0$, so g is increasing at $x = 2$.

(B) f changes concavity at $x = 1$, so it has inflection point, but at $x = 2$, there is no inflection point.

(C) f has local minimum at $x = \dfrac{1}{2}$.

(D) g' changes from decreasing to increasing at $x = \dfrac{1}{2}$, and from increasing to decreasing at $x = 2$, so g has inflection points at $x = \dfrac{1}{2}$ and $x = 2$.

84. (B)

Let $\cos(t)$ as $g(t)$.

$$f(x) = \int_0^{x^2} g(t)dt = G(x^2) - G(0)$$

$\Rightarrow f(x) = G(x^2) - G(0) \Rightarrow f'(x) = 2x \cdot g(x^2) = 2x \cdot \cos x^2$

Using the calculator, graph of $f'(x) = 2x \cdot \cos(x^2)$ can be drawn as follows.

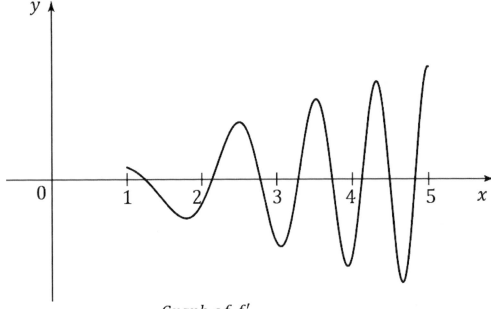

Graph of f'

That is, since there are 4 points where f' changes from negative to positive, function f has 4 relative minimum in the given interval.

85. (B)

$$\int \frac{4x^2+1-1}{1+4x^2}dx = \int \left(1-\frac{1}{1+(2x)^2}\right)dx = x - \int \frac{1}{1+(2x)^2}dx$$

Let $2x = u$, then $2 = \dfrac{du}{dx}$, so $x - \dfrac{1}{2}\int \dfrac{1}{1+u^2}du$

$\therefore x - \dfrac{1}{2}\tan^{-1}(2x) + C.$

86. (B)

I. Function f is continuous at $x = 1$. Case as follows can be made.

(Example)

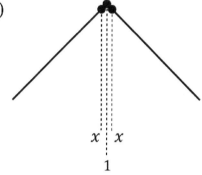

$$\lim_{x \to 1-}\frac{f(x)-f(1)}{x-1} \neq \lim_{x \to 1+}\frac{f(x)-f(1)}{x-1}$$

II. Function f is differentiable at $x = 1$, then it is continuous at $x = 1$ and $\lim\limits_{x \to 1}f(x)$ also exists.

Also, $f(1)$ exists.

III. That $f(1)$ exists doesn't always means that function f is always continuous at $x = 1$.

(Example)

87. (D)

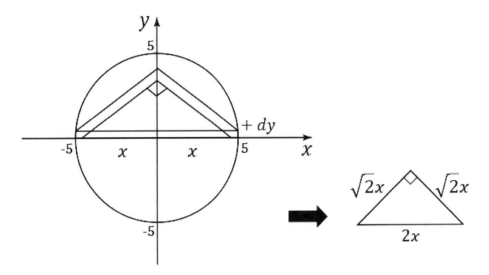

$Volume = \int (Cross\ Section\ Area) \times (Small\ Height)$

$$= \int_{-5}^{5} \frac{1}{2} \cdot \sqrt{2}\,x \cdot \sqrt{2}x \cdot dy$$

$$= \int_{-5}^{5} x^2 dy = \int_{-5}^{5} (25 - y^2) dy$$

$$= 2\int_{0}^{5} (25 - y^2) dy = 2\left[25y - \frac{1}{3}y^3\right]_{0}^{5}$$

$$= 2\left(125 - \frac{125}{3}\right) = 2 \cdot \frac{250}{3} = \frac{500}{3}$$

88. (B)

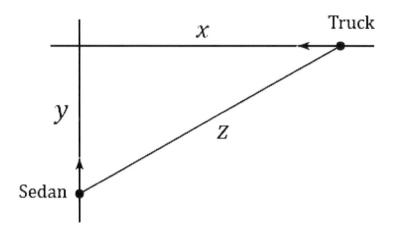

① What we have to obtain : $\dfrac{dz}{dt}$

② From $x^2 + y^2 = z^2$, $2x\dfrac{dx}{dt} + 2y\dfrac{dy}{dt} = 2z\dfrac{dz}{dt}$

③ $x = 0.05\text{km}$, $y = 0.1\text{km}$, so $z = \sqrt{0.05^2 + 0.1^2} \approx 0.1118$

Therefore, $(0.05) \times (-70) + (0.1) \times (-90) = (0.1118)\dfrac{dz}{dt}$

(* In the figure, length of x and y decreases, so $\dfrac{dx}{dt}$ and $\dfrac{dy}{dt}$ is negative.)

$\therefore \dfrac{dz}{dt} \approx -111.8$

89. (B)
$g(x) = F(x) - F(-1) \Rightarrow g'(x) = f(x)$
Therefore, the graph given can also be the graph of g'.
(A) $\lim\limits_{x \to 1-} f(x) \neq \lim\limits_{x \to 1+} f(x)$, so jump discontinuity. (False)
(B) Value of $g'(1)$ exists, so g is differentiable at $x = 1$. (True)
(C) Point where g' changes from negative to positive exists local minimum. (False)
(D) Point where f becomes concavity change exists inflection point. (False)

90. (A)

$\dfrac{dP}{dt} = kP$, so $\displaystyle\int \dfrac{1}{P} dP = \int k dt$.

$\ln|P| = kt + C \implies P = \pm e^{kt} \cdot e^c$

$\pm e^c = P_0$, so $P = P_0 e^{kt}$

(⋆ When $t = 0$, $e^C = P_0$. That is, value of C is a value when $t = 0$.)

When $t = 10$, $P = 3P_0$, and $3P_0 = P_0 e^{10k}$, so $\ln 3 = 10k$.

$k \approx 0.11$.

AP Calculus AB
Practice Test 4

A CALCULATOR MAY NOT BE USED ON THIS PART OF THE EXAMINATION.

Directions: Solve each of the following problems, using the available space for scratchwork. After examining the form of the choices, decide which is the best of the choices given and fill in the corresponding oval on the answer sheet. No credit will be given for anything written in the test book. Do not spend too much time on any one problem.

In this test:

(1) Unless otherwise specified, the domain of a function is assumed to be the set of all real numbers x for which $f(x)$ is a real number.

(B) The inverse of a trigonometric function f may be indicated using the inverse function notation f^{-1} or with the prefix "arc" (e.g., $\sin^{-1}x = \arcsin x$).

1. If $f(x) = \tan(5x)$, then $f'(x) =$

(A) $\sec(5x)$

(B) $5\sec(5x)$

(C) $5\sec^2(x)$

(D) $5\sec^2(5x)$

2. A particle moves along the x-axis so that at any time $t \geq 0$, its velocity is given by $v(t) = \cos(2t)$. If the position of the particle at time $t = \dfrac{\pi}{4}$ is $x = 1$, what is the particle's position at time $t = 0$?

(A) $-\dfrac{1}{2}$

(B) 0

(C) 1

(D) $\dfrac{1}{2}$

3. Which of the following lines is(are) asymptote(s) of the graph of $f(x) = \dfrac{x^2 - 3x - 2}{x^2 - 4}$?

I. $y = 1$

II. $x = \pm 2$

III. $x = -1$

IV. $x = -2$ and -1

(A) I only

(B) II only

(C) I and II only

(D) II and IV only

4. $\displaystyle\lim_{h \to 0} \dfrac{2(5+h)^5 - 2(5)^5}{h}$ is

(A) $f'(5)$, where $f(x) = 2x^5$

(B) $f'(5)$, where $f(x) = x^5$

(C) $f'(2)$, where $f(x) = 5x^5$

(D) $f'(2)$, where $f(x) = x^5$

5. The graph of the function $y = \frac{1}{3}x^3 + 3x^2 + 3$ changes concavity at $x=$

(A) -3

(B) -2

(C) 0

(D) 1

6. $\int_0^{\frac{\pi}{2}} \cos x e^{\sin x}\, dx =$

(A) e^2

(B) $e^2 - 1$

(C) $e^{\frac{\pi}{2}}$

(D) $e - 1$

7. If $\lim_{x\to 1} f(x) = 2$, which of the following must be true?

(A) f is continuous at $x = 1$.

(B) $f(1)$ exists.

(C) $f(x)$ is differentiable at $x = 1$.

(D) None of above.

8. Let f be a function that is continuous for all real numbers. The table below gives values of f for selected point in the closed interval [1, 12].

x	1	3	5	6	8	12
$f(x)$	2	10	2	-4	1	7

Use a trapezoidal sum with sub-intervals indicated by the data in the table to approximate $\int_1^{12} f(x)dx$.

(A) 24

(B) 28

(C) 32

(D) 36

9. The function f is twice differentiable, and the graph of f has no points of inflection. If $f(2)=2$, $f'(2)=-1$, and $f''(2)=-3$, which of the following could be the value of $f(3)$?

(A) 4

(B) 2

(C) 1

(D) 0.5

10. If $5n^2-2 < n^2 \cdot a_n < 5n^2+1$, then $\lim\limits_{n \to \infty} a_n$ is

(A) 0

(B) 5

(C) 10

(D) 25

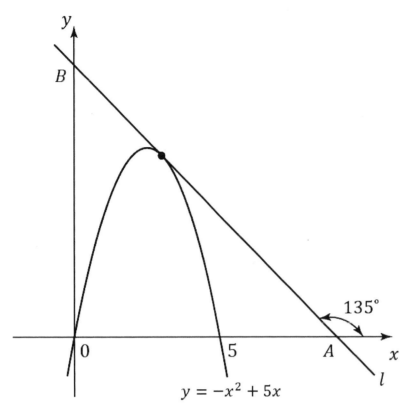

$135°$

$y = -x^2 + 5x$

11. In the figure above, line tangent l is a line that passes through a point on $y = -x^2 + 5x$. If line l is $135°$ with the positive direction of the x−axis, what is the area of the triangle OAB enclosed by the line l, x−axis and y−axis?

(A) $\dfrac{75}{2}$

(B) 39

(C) $\dfrac{81}{2}$

(D) 45

12. If $f'(x) = \dfrac{1}{x}$ and $f(e^2) = 3$, then $f(e) =$

(A) 2

(B) $\ln 3$

(C) $\ln 9$

(D) 3

13. $\displaystyle\int_{e}^{e^3} \dfrac{\ln u}{u}\,du$

(A) 1

(B) e

(C) 3

(D) 4

14. If $\lim\limits_{n\to\infty}\dfrac{\sin n\theta}{n^2}=a$ and $\lim\limits_{n\to\infty}\dfrac{5\cos n\theta}{2+n^2}=b$, then $a+b$ is

(A) $-\infty$

(B) -1

(C) 0

(D) 1

x	$f(x)$	$f'(x)$	$g(x)$	$g'(x)$
1	-2	5	-3	1
2	3	2	-2	-5
3	1	4	2	2

15. The table above gives values of f, f', g and g' at selected value of x. If $h(x)=(f\circ g)(x)$, then $h'(3)=$

(A) -4

(B) 1

(C) 4

(D) 8

16. If $f''(x) = (x-2)(x-3)^4(x-4)^2$, then the graph of $f(x)$ has inflection point(s) when $x=$

(A) 2 only

(B) 2 and 3

(C) 2 and 4

(D) 3 and 4

17. Let $f(x) = \begin{cases} \dfrac{x^2-9}{x-3} & \text{, if } x \neq 3 \\ 5 & \text{, if } x = 3 \end{cases}$, which of the following statements is(are) true?

I. $\lim\limits_{x \to 3} f(x)$ exist.

II. $f(3)$ exists.

III. f is continuous at $x = 3$.

(A) I only

(B) II only

(C) I and II only

(D) I, II and III

18. Polynomial function f satisfies following equation for all real number x.

$$f(x) = x^3 + x + \int_0^4 f(t)dt$$

What is the value of $f(1)$?

(A) -22

(B) -14

(C) 0

(D) 12

19. If $y = \dfrac{\ln(3x)}{x^2}$, then $\dfrac{dy}{dx} =$

(A) $\dfrac{3x + 2\ln(3x)}{x^3}$

(B) $\dfrac{\dfrac{1}{3}x - 2\ln(3x)}{x^3}$

(C) $\dfrac{3 - 2\ln(3x)}{x^3}$

(D) $\dfrac{1 - 2\ln(3x)}{x^3}$

20. If $f(x) = x^x$ and x is an acute angle, what is $f'(e)$?

(A) e^e

(B) $2e^e$

(C) e

(D) 0

21. If function $f(x) = (x-1)^5 + 3(x-1)^2 + 1$, what is the value of $\lim\limits_{x \to 1} \dfrac{1}{x-1} \displaystyle\int_1^{x^3} f(t)\,dt$?

(A) 1

(B) 3

(C) 6

(D) 24

22. For all real number x, $2x \cdot f'(x) + f(x) = 5x^2$. Which of the following is a quadratic function $f(x)$?

(A) x^2

(B) $x^2 + 1$

(C) $x^2 - 1$

(D) $\dfrac{1}{2}x^2$

23. If f is continuous function and if $F'(x) = f(x)$ for all real numbers x, then $\displaystyle\int_{2}^{5} f(4x)dx =$

(A) $F(20) - F(8)$

(B) $\dfrac{1}{4}(F(20) - F(8))$

(C) $4(F(20) - F(8))$

(D) $\dfrac{1}{4}(F(5) - F(2))$

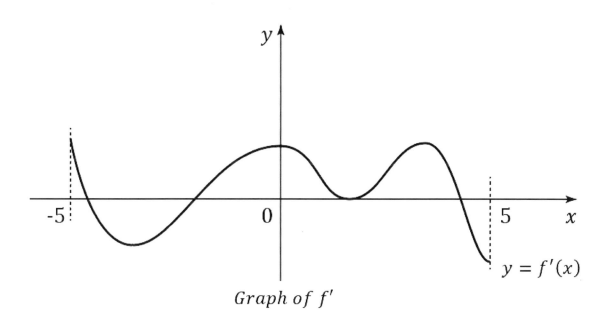

Graph of f'

24. The figure above shows the graph of f', the derivative of the function f, on the interval $[-5, 5]$. If the derivative of the function g is given by $g'(x) = 3f'(x)$, how many points of inflection does the graph of g have on the interval $[-5, 5]$?

(A) One

(B) Two

(C) Three

(D) Four

25. Let f be the function defined by $f(x) = \dfrac{1}{3}x^3 - 2x^2 + 3x + 2$. What is the intervals of f both decreasing and concave up?

(A) $x > 1$

(B) $2 < x < 3$

(C) $1 < x < 3$

(D) $x > 3$

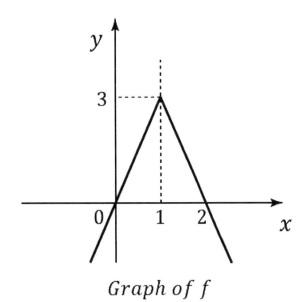

Graph of f

26. Figure above is the graph of function f. What is the value of $\displaystyle\int_{-1}^{3} xf(x)dx$?

(A) 0

(B) 2

(C) 4

(D) 8

27. If f is the function given by $f(x) = \int_{1}^{2x} \sqrt{t^3 + t} \, dt$, then $f'(1) =$

(A) $2\sqrt{10}$

(B) 3

(C) $\sqrt{10}$

(D) $\sqrt{12}$

28. Two particles, P and Q start simultaneously on the $x-$ axis and moving in the positive direction of the $x-$axis. Let differentiable function $P(t)$ and $Q(t)$ denote the distance that two particles move for t seconds, then $P(t)$ and $Q(t)$ satisfy the following conditions.

> I. $P(5) = Q(5)$
>
> II. $P'(t) < Q'(t)$ when $1 \le t \le 10$.

When $1 \le t \le 10$, which of the following is true about the position of two particles P and Q?

(A) Particle P is always ahead of particle Q.

(B) Particle Q is always ahead of particle P.

(C) Particle P overtakes particle Q once.

(D) Particle Q overtakes particle P once.

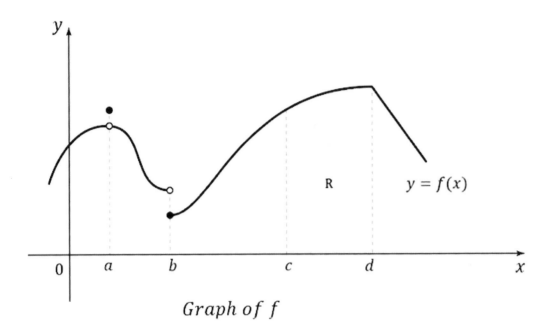

Graph of f

29. The graph above is the graph of $y = f(x)$. Which of the following must be true?

(A) f is jump discontinuous at $x = a$.

(B) $\lim\limits_{x \to b-} f(x) = \lim\limits_{x \to b+} f(x)$

(C) $\lim\limits_{x \to d} f(x)$ does not exist.

(D) Area $R = \lim\limits_{n \to \infty} \sum\limits_{k=1}^{n} f(c + \dfrac{d-c}{n} \cdot k) \cdot \dfrac{d-c}{n}$

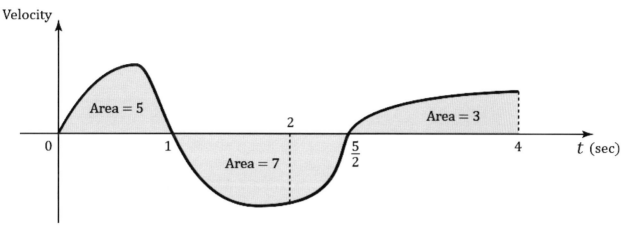

30. The differentiable graph above is the graph about the velocity of an object that moves along $x-$axis after t seconds. Which of the following must be true?

(A) Let the position of the object at time $t=0$ as $x(0)=3$, then $x(1)=5$.

(B) At time $t=\dfrac{5}{2}$, this object locates on the farthest to the right.

(C) At time $t=2$, the speed is increasing.

(D) Let the position of this object after c seconds as $x(c)$ and $x(0)=3$, then at $1\leq c\leq \dfrac{5}{2}$, at least one c exists that satisfies $x(c)=5$.

CALCULUS AB
SECTION I, PART B
Time-45 minutes
Number of questions-15

A GRAPHING CALCULATOR IS REQUIRED FOR SOME QUESTIONS ON THIS PART OF THE EXAMINATION.

Directions: Solve each of the following problems, using the available space for scratchwork. After examining the form of the choices, decide which is the best of choices given and fill in the corresponding oval on the answer sheet. No credit will be given for anything written in the test book. Do no spend too much time on any one problem.

BE SURE YOU ARE USING PAGE 3 OF THE ANSWER SHEET TO RECORD YOUR ANSWERS TO QUESTIONS NUMBERED 76-90.

YOU MAY NOT RETURN TO PAGE 2 OF THE ANSWER SHEET.

In this test:

(1) The exact numerical value of the correct does not always appear among the choices given. When this happens, select from among the choices the number that best approximates the exact numerical value.

(2) Unless otherwise specified, the domain of a function f is assumed to be the set of all real numbers x for which $f(x)$ is real number.

(3) The inverse of a trigonometric function f may be indicated using the inverse function notation f^{-1} or with the prefix "arc" (e.g, $\sin^{-1}x = \arcsin x$)

76. If $\displaystyle\int_0^2 f(x)dx = -3$ and $\displaystyle\int_{-1}^0 f(x)dx = 5$, then $\displaystyle\int_{-1}^2 (f(x)+2)dx =$

(A) -4

(B) -2

(C) 4

(D) 8

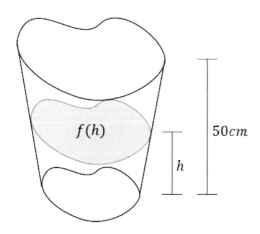

77. Pour water into a 50cm tall jar as shown above. Let $f(h)$ be the surface area of the water. What is the expression of the volume of water in the jar until the height of the water from the bottom of the jar is 20cm?

(A) $f(20)$

(B) $f'(20)-f'(0)$

(C) $\displaystyle\int_0^{20} f(h)dh$

(D) $\displaystyle\int_0^{20} f'(h)dh$

78. Suppose $f(x) = \begin{cases} \dfrac{5x(x-1)}{x^2 - 5x + 4}, & \text{for } x \neq 1, 4 \\ -\dfrac{5}{3}, & \text{for } x = 1 \\ 0, & \text{for } x = 4 \end{cases}$, and $f(x)$ is continuous

(A) except at $x = 1$

(B) except at $x = 4$

(C) except at $x = 1$ or $x = 4$

(D) for all real number

79. When the region enclosed by the graphs of $y = x$ and $y = 5x - x^2$ is revolved about the y-axis, the volume of the solid generated is given by

(A) $2\pi \displaystyle\int_0^4 (4x - x^2)\, dx$

(B) $\pi \displaystyle\int_0^4 (4x - x^2)\, dx$

(C) $2\pi \displaystyle\int_0^4 (4x^2 - x^3)\, dx$

(D) $\pi \displaystyle\int_0^4 (4x^2 - x^3)\, dx$

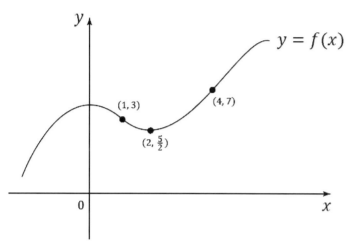

80. The graph above is the graph of $y = f(x)$ which is a differentiable function. $g(x) = \dfrac{d}{dx} \displaystyle\int_0^{x^2} f(t)dt$, then what is the value of $g(1)$?

(A) 3

(B) $\dfrac{5}{2}$

(C) 6

(D) 7

81. The base of a solid is the region bounded by $y = e^x$, $x = 1$, $x = 3$ and x-axis. Each cross section perpendicular to the x-axis is a square. What is the volume of solid?

(A) 52.31

(B) 88.35

(C) 126.21

(D) 198.02

x	1	2	3	4	5	6
$f(x)$	5	1	-1	0	-3	5

82. The function f is continuous and differentiable on the close interval $[1,6]$. The table above gives selected values of f on this interval. Which of the following statements must be true?

(A) f have three inflection points.

(B) The maximum value of f on $[1,6]$ is 5.

(C) There is a number of c in the open interval $(1,6)$ such that $f'(c)=0$.

(D) $f(x)>0$ for $1<x<2$.

83. $\dfrac{d}{dx}\displaystyle\int_{10}^{\sqrt{x}}\ln(t^2+1)dt =$

(A) $\ln(x+1)$

(B) $\dfrac{1}{2\sqrt{x}}\ln(x^2+1)$

(C) $\dfrac{1}{2\sqrt{x}}\ln(x+1)$

(D) $\ln(\sqrt{x}+1)$

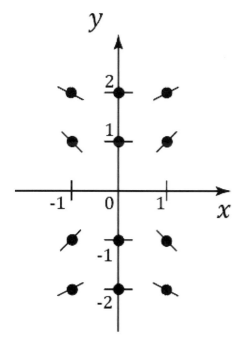

84. Shown above is the slope field for which of the following differential equations?

(A) $\dfrac{dy}{dx} = xy$

(B) $\dfrac{dy}{dx} = x - y$

(C) $\dfrac{dy}{dx} = \dfrac{y}{x}$

(D) $\dfrac{dy}{dx} = \dfrac{x}{y}$

85. Polynomial function $f(x)$ satisfies the following equation.

$$\int_1^x (x-t)f(t)dt = x^3 + ax^2 + x + b$$

Which of the following is the value of $2a+3b$? (x is all real number.)

(A) -10

(B) -4

(C) 0

(D) 12

86. $\dfrac{d}{dx}\displaystyle\int_0^x \sin(5t)dt=$

(A) $\sin(5x)$

(B) $5\sin(5x)$

(C) $\cos(5x)$

(D) $\dfrac{1}{5}\sin(5x)$

87. If g is the inverse function of f and if $f(x) = -x^3$, then $g'(1)=$

(A) $-\dfrac{1}{3}$

(B) -3

(C) 1

(D) $\dfrac{1}{3}$

88. The function f has the property that $f'(x) > 0$, and $f''(x) < 0$ for all real values x. Which of the following could be graph of f?

(A)

(B)

(C)

(D)

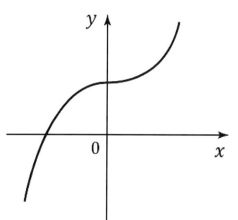

89. At time $t=0$, pizza with temperature $95\,^\circ\text{C}$ was removed from an oven. The temperature of the pizza at time t is modeled by a differentiable function P for which it is known that $\dfrac{dP}{dt}=-7.4e^{-0.13t}$. What is the temperature of the pizza at time $t=10$?

(A) 37.25

(B) 41.41

(C) 53.59

(D) 60.32

90. Function $f(x)=2xe^{ax+b}$ satisfies $f'(0)=1$ and $f''(0)=2$. What is the value of $a+b$?

(A) 1

(B) 2

(C) $1+\ln\dfrac{1}{2}$

(D) $2+\ln 2$

AP Calculus AB Practice Test 4
❯ Answer Key

Part A

1. (D)	6. (D)	11. (C)	16. (A)	21. (A)	26. (A)
2. (D)	7. (D)	12. (A)	17. (C)	22. (C)	27. (A)
3. (C)	8. (D)	13. (D)	18. (A)	23. (B)	28. (D)
4. (A)	9. (D)	14. (C)	19. (D)	24. (D)	29. (D)
5. (A)	10. (B)	15. (C)	20. (B)	25. (B)	30. (D)

Part B

76. (D)	79. (C)	82. (C)	85. (B)	88. (C)
77. (C)	80. (C)	83. (C)	86. (A)	89. (C)
78. (B)	81. (D)	84. (D)	87. (A)	90. (C)

PART A

1. (D)

$f'(x) = 5\sec^2(5x)$

2. (D)

When the Position from $t=0$ is $x(0)$,

$$x(0) = x\left(\frac{\pi}{4}\right) + \int_{\frac{\pi}{4}}^{0} \cos(2t)dt = x\left(\frac{\pi}{4}\right) - \int_{0}^{\frac{\pi}{4}} \cos(2t)dt$$

$$\Rightarrow 2t = u, \ 2 = \frac{du}{dt}$$

$$\Rightarrow 1 - \frac{1}{2}\int_{0}^{\frac{\pi}{2}} \cos u\, du$$

$$= 1 - \frac{1}{2}[\sin u]_{0}^{\frac{\pi}{2}} = 1 - \frac{1}{2} = \frac{1}{2}$$

3. (C)

Horizontal line: $\lim_{x \to \pm\infty} f(x) = 1 \Rightarrow y = 1$

Vertical line: $x^2 - 4 = 0 \Rightarrow x = \pm 2$

4. (A)

Since $\lim_{h \to 0} \dfrac{f(x+h) - f(x)}{h} = f'(x)$, $x = 5$ and $f(x) = 2x^5$.

Therefore, when $f(x) = 2x^5$, $f'(5)$.

284

5. (A)

......changes concavity........ = Inflection Point

Find the point where y' changes from increasing to decreasing, or from decreasing to increasing,

or

find the point where the sign change of y'' happens.

$\Rightarrow y' = x^2 + 6x$

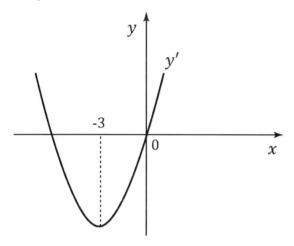

$\Rightarrow y'' = 2x + 6$

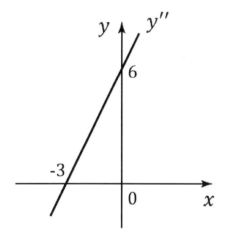

Therefore, the graph changes concavity at $x = -3$.

6. (D)

$\sin x = u \Rightarrow \cos x = \dfrac{du}{dx}$

$\Rightarrow \displaystyle\int_0^1 \cos x\, e^u \frac{1}{\cos x} du = \int_0^1 e^u du = [e^u]_0^1 = e - 1$

7. (D)

$\lim_{x \to 1} f(x) = 2$ means that $\lim_{x \to 1} f(x)$ exists. Graph of $y = f(x)$ that $\lim_{x \to 1} f(x)$ exists can be as follows.

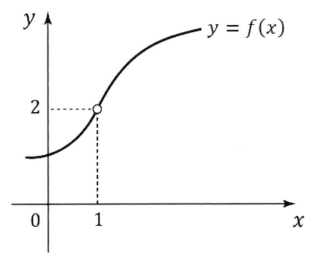

That is, even though $\lim_{x \to 1^-} f(x) = \lim_{x \to 1^+} f(x) = 2$, $f(1)$ can exist or not like in the figure above, and can be discontinuous and not differentiable at $x = 1$.

8. (D)

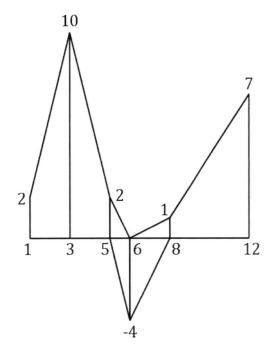

$$T = \frac{1}{2}[(2+10)\times 2 + (10+2)\times 2 + (2-4) + (-4+1)\times 2 + (1+7)\times 4] = 36$$

9. (D)

- $f'(2) = -1 < 0 \Rightarrow$ Decreasing!
- $f''(2) = -3 < 0 \Rightarrow$ Concave Down!
- "... has no points of inflection..." $\Rightarrow$ concavity doesn't change!

Situation given can be drawn as follows.

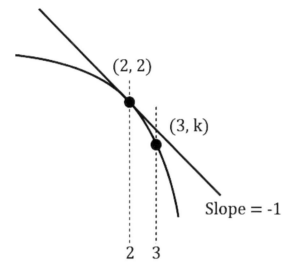

Using Mean Value Theorem, $f'(2.5) = \dfrac{k-2}{3-2} < -1$ should be satisfied.

10. (B)

Divide both sides with n^2 and $\lim\limits_{n \to \infty}$, then

From $\lim\limits_{n \to \infty} \dfrac{5n^2 - 2}{n^2} \leq \lim\limits_{n \to \infty} a_n \leq \lim\limits_{n \to \infty} \dfrac{5n^2 + 1}{n^2}$, $5 \leq \lim\limits_{n \to \infty} a_n \leq 5$, therefore $\lim\limits_{n \to \infty} a_n = 5$.

※ Both $\lim\limits_{n \to \infty} \dfrac{5n^2 - 2}{n^2} = 5$ and $\lim\limits_{n \to \infty} \dfrac{5n^2 + 1}{n^2} = 5$ are not exact values, just approximation,

therefore $\lim\limits_{n \to \infty} a_n = 5$ is held.

11. (C)

Slope of line $\ell = \tan 135° = -1$, therefore from $y' = -2x + 5 = -1$,

$x = 3$ and $y = -3^2 + 5 \cdot 3 = 6$.

Therefore, equation of line tangent ℓ is which has -1 as a slope and passes point $(3, 6)$.

Therefore, from $y - 6 = -(x - 3)$, $y = -x + 9$, therefore $A(9, 0)$ and $B(0, 9)$.

Therefore, area of $\triangle OAB$ is $\dfrac{1}{2} \cdot 9 \cdot 9 = \dfrac{81}{2}$.

12. (A)

$f(x) = \displaystyle\int \dfrac{1}{x} dx = \ln|x| + C \Rightarrow 3 = \ln e^2 + C,\ C = 1$

$f(x) = \ln|x| + 1 \Rightarrow f(e) = 2$

13. (D)

$\ln u = x,\ \dfrac{1}{u} = \dfrac{dx}{du}$

$\displaystyle\int_1^3 x \times \dfrac{1}{u} \times u\, dx = \int_1^3 x\, dx = [\dfrac{1}{2}x^2]_1^3 = 4$

14. (C)

From $\displaystyle\lim_{n \to \infty} \dfrac{\sin n\theta}{n^2}$, $\displaystyle\lim_{n \to \infty} \dfrac{-1 \sim 1}{n^2} = 0 = a$

From $\displaystyle\lim_{n \to \infty} \dfrac{5\cos n\theta}{2 + n^2}$, $\displaystyle\lim_{n \to \infty} \dfrac{-5 \sim 5}{n^2 + 2} = 0 = b$

$\therefore a + b = 0$

15. (C)

$h'(x) = f'(g(x)) \times g'(x) \Rightarrow h'(3) = f'(g(3)) \times g'(3) = 2 \times 2 = 4$

16. (A)

$f''(x) = 0$ and f'' changes sign at $x = 2$

(※ Inflection point is where sign change of f'' happens.)

17. (C)

I. $\lim\limits_{x \to 3-} f(x) = \lim\limits_{x \to 3+} f(x) = 6$

II. $f(3) = 5$, Therefore, $f(3)$ exist.

III. $\lim\limits_{x \to 3} \dfrac{x^2 - 9}{x - 3} = 6 \neq f(3)$. Therefore, f is discontinuous at $x = 3$.

18. (A)

Since $\displaystyle\int_0^4 f(t)dt$ is a constant, let $\displaystyle\int_0^4 f(t)dt = C$, then $f(x) = x^3 + x + C$.

From $C = \displaystyle\int_0^4 (x^3 + x + C)dx$, $C = \left[\dfrac{1}{4}x^4 + \dfrac{1}{2}x^2 + Cx \right]_0^4 = \dfrac{1}{4} \cdot 4^4 + \dfrac{1}{2} \cdot 4^2 + 4C$.

From $C = 4^3 + 8 + 4C$, $-3C = 72$, therefore $C = -24$.

Therefore, $f(x) = x^2 + x - 24$, $f(1) = -22$.

19. (D)

$$\frac{dy}{dx} = \frac{\dfrac{3}{3x} \times x^2 - \ln(3x) \times 2x}{x^4} = \frac{1 - 2\ln(3x)}{x^3}$$

20. (B)

Put ln on both sides, then from $\ln f(x) = x \ln x$, $\dfrac{f'(x)}{f(x)} = \ln x + x \cdot \dfrac{1}{x}$

$\Rightarrow f'(x) = f(x)\{\ln x + 1\}$, therefore $f'(x) = x^x(\ln x + 1)$.

Therefore, $f'(e) = 2e^e$

21. (B)

From $\displaystyle\int_1^{x^3} f(t)dt = F(x^3) - F(1)$, $\displaystyle\lim_{x \to 1}\frac{F(x^3) - F(1)}{x - 1}$ is the form of $\dfrac{0}{0}$,

therefore using L'Hopital's Rule, $\displaystyle\lim_{x \to 1} 3x^2 f(x^3) = 3f(1)$.

Therefore, $3f(1) = 3$.

22. (A)

Let $f(x) = ax^2 + bx + c$, then $f'(x) = 2ax + b$.

Substitute this in $2xf'(x) + f(x) = 5x^2$, then from $2x(2ax + b) + ax^2 + bx + c = 5x^2$,

$5ax^2 + 3bx + c = 5x^2$, therefore $5a = 5$, $3b = 0$, $c = 0$.

We get $a = 1$, $b = 0$, $c = 0$.

Therefore, $f(x) = x^2$.

23. (B)

$4x = u, \; 4 = \dfrac{du}{dx}$

$\dfrac{1}{4}\displaystyle\int_{8}^{20} f(u)du = \dfrac{1}{4}[F(20) - F(8)]$

24. (D)

Inflection point is where graph of g' changes from increasing to decreasing, or from decreasing to increasing.

Since $g'(x) = 3f'(x)$, the given graph increases 3 times to $y-$axis, but the value of x where the inflection point is made does not change.

Therefore, g has 4 inflection points.

25. (B)

- $f'(x) = x^2 - 4x + 3 < 0 \;\Rightarrow\;$ Decreasing!

  $\Rightarrow\; 1 < x < 3$

- $f''(x) = 2x - 4 > 0 \;\Rightarrow\;$ Concave up

  $\Rightarrow\; x > 2$

Therefore, interval where function f is both decreasing and concave up is $2 < x < 3$.

26. (A)

It is $f(x) = -3x + 6$ when $x \geq 1$, and $f(x) = 3x$ when $x < 1$.

Therefore, $\displaystyle\int_{-1}^{3} xf(x)dx = \int_{-1}^{1} 3x^2 dx + \int_{1}^{3} (-3x^2 + 6x)dx$

$= [x^3]_{-1}^{1} + [-x^3 + 3x^2]_{1}^{3} = 2 + \{(-27+27) - (-1+3)\} = 0$

27. (A)

$\sqrt{t^3 + t} = g(t)$

$f(x) = \displaystyle\int_{1}^{2x} g(t)dt = G(2x) - G(1) \Rightarrow f'(x) = 2g(2x) = 2\sqrt{8x^3 + 2x}$

$\therefore f'(1) = 2\sqrt{10}$

28. (D)

I. When time $t = 5$, two particles are in the same location.

II. When $1 \leq t \leq 10$, $P'(t) < Q'(t)$, therefore the velocity of particle Q is faster than the velocity of particle P.

That is, at $1 \leq t < 5$, particle P is ahead of particle Q, and when time $t = 5$, two particles are in the same location. At $5 < t \leq 10$, particle Q overtakes particle P. Therefore, particle Q overtakes particle P once.

29. (D)

(A) f is removable discontinuity at $x = a$ because $\lim\limits_{x \to a} f(x)$ exists.

(B) $\lim\limits_{x \to b-} f(x) \neq \lim\limits_{x \to b+} f(x)$

(C) $\lim\limits_{x \to d-} f(x) = \lim\limits_{x \to d+} f(x)$. Therefore $\lim\limits_{x \to d} f(x)$ exists.

(D)

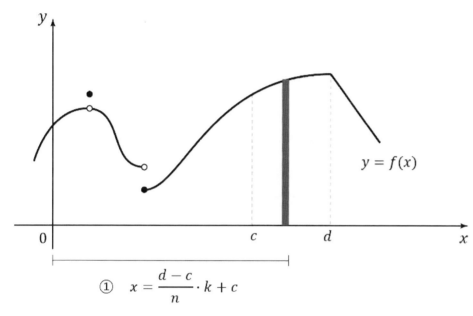

① $x = \dfrac{d-c}{n} \cdot k + c$

② $\dfrac{d-c}{n} = dx$

Therefore, area $R = \lim\limits_{n \to \infty} \sum\limits_{k=1}^{n} f\left(c + \dfrac{d-c}{n} \cdot k\right) \cdot \dfrac{d-c}{n}$

30. (D)

(A) Let the position of the object at $t = 0$ as $x(0) = 3$,

$x(1) - x(0) = \int_0^1 v(t)dt = 5$, so $x(1) = 8$.

(B) $x(1) = 8$.

$x(\frac{5}{2}) - x(1) = \int_1^{\frac{5}{2}} v(t)dt$, so $x(\frac{5}{2}) = 8 - 7 = 1$.

$x(4) - x(\frac{5}{2}) = \int_{\frac{5}{2}}^4 v(t)dt$, so $x(4) = 1 + \int_{\frac{5}{2}}^4 v(t)dt = 4$.

Therefore, at $t = 1$, the object is at the far right.

(C) At $t = 2$, velocity is negative, and acceleration is positive.

Therefore, the speed is decreasing.

(D) Position at $t = 1$ is $x(1) = 8$, and the position at $t = \frac{5}{2}$ is $x(\frac{5}{2}) = 1$.

Position is continuous, so by Intermediate Value Theorem,

in the interval $1 \le t \le \frac{5}{2}$, at least one c exists that satisfies $x(c) = 5$.

PART B

76. (D)

$$\int_0^2 f(x)dx + \int_{-1}^0 f(x)dx = \int_{-1}^2 f(x)dx = 2$$

$$\int_{-1}^2 (f(x)+2)dx = \int_{-1}^2 f(x)dx + \int_{-1}^2 2dx = 2 + [2x]_{-1}^2 = 2 + (4+2) = 8$$

77. (C)

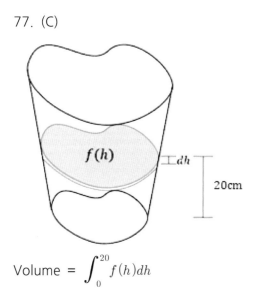

Volume = $\int_0^{20} f(h)dh$

78. (B)

$$\lim_{x \to 1} \frac{5x(x-1)}{x^2 - 5x + 4} = -\frac{5}{3} = f(1)$$

$$\lim_{x \to 4} \frac{5x(x-1)}{x^2 - 5x + 4} = \pm\infty \neq f(4)$$

Thus, f is discontinuous at $x = 4$

79. (C)

Shell Method

$$\text{Volume}=2\pi\int_a^b xy\,dx = 2\pi\int_0^4 xy\,dx$$

$$2\pi\int_0^4 x(5x - x^2 - x)\,dx = 2\pi\int_0^4 (4x^2 - x^3)\,dx$$

80. (C)

$$\frac{d}{dx}(F(x^2) - F(0)) = 2xf(x^2)$$

Therefore, from $g(x) = 2xf(x^2)$, $g(1) = 2f(1) = 6$.

81. (D)

$$\text{Volume}= \int_1^3 (\text{Cross Section Area})\,dx \Rightarrow \text{Volume}= \int_1^3 e^{2x}\,dx \approx 198.02$$

82. (C)

The condition of the problem is Differential, and $f(1) = f(6)$, using the Rolle's Theorem. Thus, at least one exists from $f'(c) = 0$ within the interval, where c is given.

83. (C)

Let $\ln(t^2 + t) = f(t)$.

$$\int_{10}^{\sqrt{x}} f(t)\,dt = F(\sqrt{x}) - F(10), \text{ so } \frac{d}{dx}(F(\sqrt{x}) - F(10)) = \frac{1}{2\sqrt{x}}f(\sqrt{x})$$

Therefore, $\dfrac{1}{2\sqrt{x}}\ln(x+1)$.

84. (D)

Substitute $(1, 1)$, $(1, 2)$, $(1, -1)$, $(1, -2)$ $\cdots$, then $\dfrac{dy}{dx} = \dfrac{x}{y}$ is the answer.

85. (B)

$\displaystyle\int_1^x (x-t)f(t)dt = x^3 + ax^2 + x + b$

① $x = 1$, $1 + a + 1 + b = 0$ $\therefore a + b = -2$

② $x\displaystyle\int_1^x f(t)dt - \int_1^x tf(t)dt = x^3 + ax^2 + x + b$. Let $tf(t) = g(t)$. Then, $g(1) = f(1)$.

$x(F(x) - F(1)) - (G(x) - G(1)) = x^3 + ax^2 + x + b$ or $xF(x) - xF(1) - (G(x) - G(1)) = x^3 + ax^2 + x + b$

$\Rightarrow F(x) + xf(x) - F(1) - g(x) = 3x^2 + 2ax + 1$

$\Rightarrow F(1) + f(1) - F(1) - g(1) = 3 + 2a + 1$

$\Rightarrow 0 = 3 + 2a + 1$

$\therefore a = -2$, $b = 0$

86. (A)

$f(t) = \sin(5t)$

$\dfrac{d}{dx}\displaystyle\int_0^x f(t)dt = \dfrac{d}{dt}(F(x) - F(0)) = f(x) = \sin(5x)$

87. (A)

$g'(1) = (f^{-1})'(1) = \dfrac{1}{f'(-1)}$ $(※ 1 = -x^3 \Rightarrow x = -1)$

$f'(x) = -3x^2 \Rightarrow f'(-1) = -3$

$\therefore g'(1) = \dfrac{1}{f'(-1)} = -\dfrac{1}{3}$

88. (C)

① $f'(x) > 0 \Rightarrow f$ is increasing

② $f''(x) < 0 \Rightarrow f$ is concave down

Therefore, the answer is (C).

89. (C)

From $P'(t) = \dfrac{dP}{dt} = -7.4e^{-0.13t}$, $P(10) - P(0) = \displaystyle\int_0^{10} P'(t)dt$

$\Rightarrow P(10) = 95 + \displaystyle\int_0^{10} P'(t)dt \approx 53.59$

90. (C)

① In $f'(x) = 2e^{ax+b} + 2axe^{ax+b} \cdot 2$, $f'(0) = 2e^b = 1$, so $e^b = \dfrac{1}{2}$.

Therefore $b = \ln\dfrac{1}{2}$.

② In $f''(x) = 2ae^{ax+b} + 2ae^{ax+b} + 2ax \cdot e^{ax+b} \cdot a$, $f''(0) = 2ae^b + 2ae^b = 2$, so $4ae^b = 2$.

In $ae^b = \dfrac{1}{2}$, $b = \ln\dfrac{1}{2}$, so from $ae^{\ln\frac{1}{2}} = \dfrac{1}{2}$, $a = 1$.

Therefore, $a + b = 1 + \ln\dfrac{1}{2}$.

AP Calculus AB
Free Response Questions
 6 Topics로 만점 받기

 시작하기 전에 …

AP Calculus는 무조건 다 맞아야 만점이 나오는 시험이 아니다. 어느 정도까지 풀어내면 만점이 나오는 시험이다. AP Calculus AB를 6개의 Topic으로 총정리하였다. 여기에 나오는 Topic들만 정확히 알고 시험장에 들어가도 충분히 5점 만점이 나오리라 확신하는 바이다.

6개의 Topic을 모두 공부하였다면 www.collegeboard.org에 공개되어있는
Free Response Question 최근 문제 3~5년 정도의 분량을 더 풀어보기 바란다.

반드시 공개되어있는 답처럼 쓰지 않아도 된다. 여기에서 제시하는 풀이가 Collegeboard에서 제시하는 풀이 방식과 다르다고 하여 걱정할 필요는 없다.

심선생의 주절주절....1

필자는 3월만 되면 몹시 긴장이 된다. 3월에는 AIME시험도 있고 5월에 AP Calculus AB와 BC 시험도 있기 때문이다. 필자의 수업을 들어오는 학생들을 만날때마다 항상 겪어왔던 어려운점은 바로 과제를 마음대로 낼 수 없다는 점이었다. 한국에서 수학을 가르칠때는 과제를 필자 마음대로 냈었지만 유학을 준비하는 학생들은 그럴 수가 없었다. 이유는 수학이 대학으로 가기위한 첫 번째가 아닌 점도 있지만 공부 이외에도 바쁜 일들이 너무 많아 보이기도 했기 때문이다. 과제를 내더라도 제대로 해오는 학생들은 20%미만 이었다.

 학생들이나 부모님들이나 원하는 것은 모두 똑 같았다.
"과제는 제로!! 그래도 점수는 Perfect!!"
도둑 심보 같지만 사실 충분히 이해가 가는 부분이기도 하다. 오래전 일이지만 필자는 AP Calculus AB와 BC 시험에 직접 응시를 했었다. 무슨 문제가 나올지는 궁금하지 않았다. 어짜피 어느 정도 문제들이 공개가 되기에...정말 궁금했던 것은 점수를 어떻게 계산하는지와 어떻게 하면 4점이 나오는지를 테스트해보고 싶은 것이었다.

놀랐던 사실은....4점 이라고 생각했던 성적이 계속 5점이 나왔다는 사실이었다. 오래전 일이기는 하지만 이와 같은 일은 지금도 계속되고 있다. 학생들의 말을 통해서 이는 계속 확인이 되고 있는 상황이다.

 그 이후 필자는 수업자료를 반 이상 줄여서 수업을 진행하였으며 과제도 거의 안내는 방향으로 수업을 진행하였다. AP Calculus와는 다르게 History나 Chemistry, Biology 등은 적당히 답을 써서는 만점을 받을 수가 없고 공부하는 분량 또한 AP Calculus에 비해 상당히 많은데 많은 학생들이 AP시험을 여러과목을 치르기에 AP Calculus 이라도 시간을 절약해줘야 한다는 책임감이 들어서였다. 첫 한 두해 동안은 솔직히 필자도 걱정이 되기는 하였으나 시간이 지날수록 이것이 맞는 방법이라는 확신이 들게 되었다. 대부분의 학생들이 적은 과제량으로도 만점을 받는 것을 확인해왔기 때문이다.

 한국에서 공부를 하였던 학생들은 수능처럼 생각을 하여 수학에 더 큰 비중을 두고 공부하다보니 History, Chemistry, Biology에 비해 AP Calculus에 더 많은 시간을 들이는 학생들을 자주 보게 되는데 이는 효과적인 시험 준비가 아니라고 말씀드리고 싶다.

 성공하는 사람은 시간을 잘 활용하는 사람이다. 명문대에 진학한 학생들을 보면 천재라기 보다는 시간 계획을 잘 세웠던 학생들이 많았다. 종종 이런 부모님들께서 이런 말씀을 하신다. "우리 아이는 SAT준비 때문에 시간이 없어요...." 이는 누구나 마찬가지이다. 하지만 명문대 진학한 학생들의 입에서 필자는 이런 말을 거의 들어본 적이 없다. 대부분 적당히 준비하여 점수를 잘 받은 학생들이 많았는데 이는 바쁜 학기중에도 시간을 내서 틈틈이 공부를 하면서 시간을 절약해 왔기 때문이다. 이런 말이 있지 않은가?-"시간은 금이다!!"

Topic 1

Graph

다음의 $y = f(x)$, $y = f'(x)$, $y = f''(x)$의 graph들을 보고 그 내용들을 암기하자.

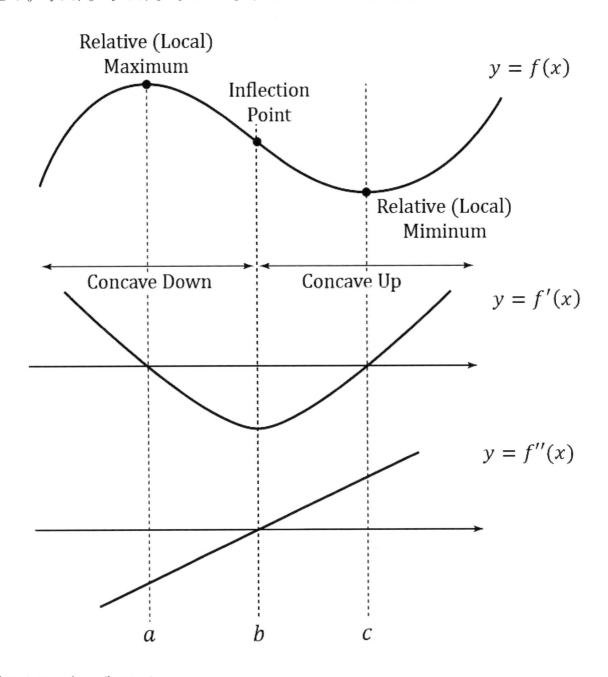

① Relative (Local) Maximum

① f'가 positive에서 negative로 change 되는 point
② $f'(x) = 0$이면서 $f''(x) < 0$인 point

② Relative (Local) Minimum

① f'이 negative에서 positive로 change 되는 point
② $f'(x) = 0$이면서 $f''(x) > 0$인 point

※Relative Maximum or Minimum을 찾을 때 $f'(x) = 0$인 point라고 하면 안 된다. 반드시 $f''(x) < 0$ or $f''(x) > 0$인 조건이 같이 들어가야 한다.

③ Inflection Point

① f'이 increasing에서 decreasing으로 또는 decreasing에서 increasing으로 change 되는 point
② f''이 sign change가 생기는 point

※Inflection point를 구할 때 $f''(x)=0$인 점이라고 하면 안 된다.

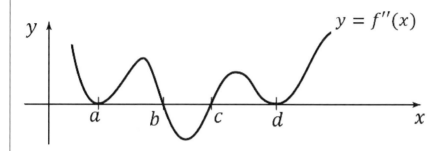

위의 $y=f''(x)$로부터 알 수 있는 것은 $y=f(x)$는 inflection point를 $x=b$와 $x=c$에서 갖는다는 것이다. $x=a$, $x=d$에서는 f''의 sign change가 일어나지 않기 때문에 inflection point가 생기지 않는다.

---

④ Concave Down

① $y=f'(x)$이 decreasing하는 구간에서 발생
② $f''(x)<0$인 구간에서 발생

⑤ Concave Up

① $y=f'(x)$이 increasing하는 구간에서 발생
② $f''(x)>0$인 구간에서 발생

⑥ Increasing

① $f'(x)>0$인 구간에서 발생

⑦ Decreasing

① $f'(x)<0$인 구간에서 발생

위의 7가지는 반드시 암기하기 바란다.

Graph 문제를 해결하는 데 있어서 다음의 사항을 반드시 알아두어야 한다. 대부분의 학생들이 어려워하는 내용이다.

1. $\int$ 은 Graph를 해석하는 데 도움이 안 된다.

Definite Integral의 definition을 이용하여 다음과 같이 제거한다.

① $g(x) = \displaystyle\int_a^x f(t)dt$

$\Rightarrow$ ② $g(x) = F(x) - F(a)$

$\Rightarrow$ ③ $g'(x) = f(x)$

※ 여기에서 $F(a)$는 Constant이므로 $F'(a)$는 무조건 0이 된다.

2. Absolute Maximum Value와 Absolute Minimum Value는 다음과 같이 구하자.

① f' graph로부터 f graph를 추정하여 Maximum Value와 Minimum Value 후보를 선정한다.

② 예를 들어, Maximum Value의 후보가 $f(b)$와 $f(a)$라면

$f(b) - f(a) = \displaystyle\int_a^b f'(x)dx$를 이용하여 $\displaystyle\int_a^b f'(x)dx > 0$인지 또는 $\displaystyle\int_a^b f'(x)dx < 0$인지를 조사한다.

※ (1) 3이 1보다 크다는 사실은 누구나 아는 사실이다. 이를 증명하는 방법은 $3 - 1 > 0 \Rightarrow 3 > 1$. 즉, 두 값을 빼 보면 알게 된다. 위의 ②에서 설명한 것도 이와 같은 원리이다.

(2) $\displaystyle\int_a^b f'(x)dx = f(b) - f(a)$인 것은 쉽게 알 수 있지만 이를 거꾸로 쓰는 $f(b) - f(a) = \displaystyle\int_a^b f'(x)dx$는 생각보다 잘 써지지 않는다.

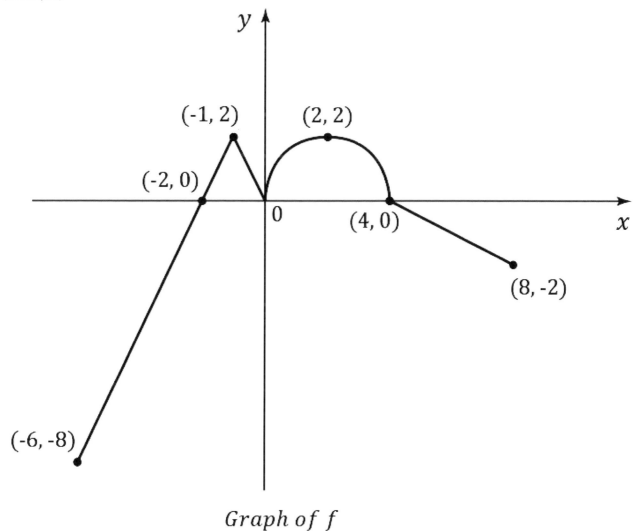

Graph of f

The graph of the function f shown above consists of a semicircle and three line segments. Let g be the function given by $g(x) = \int_{-1}^{x} f(t)dt$.

(1) Find all values of x in the open interval $(-6, 8)$ at which g attains a local minimum. Justify your answer.

(2) Find all values of x in the open interval $(-6, 8)$ at which the graph of g has a point of inflection.

(3) Find the absolute minimum value of g on the closed interval $[-6, 8]$. Justify your answer.

(Solution)

$g(x) = \displaystyle\int_{-1}^{x} f(t)dt = F(x) - F(-1)$, therefore $g'(x) - f(x)$.

That is, the graph given is both f graph and g' graph.

(1) Local Minimum is where g changes from negative to positive, therefore the answer is $x = -2$.

$\Rightarrow g' = f$ changes from negative to positive at $x = -2$.

(2) Inflection point is where g' graph changes from increasing to decreasing or from decreasing to increasing, therefore the answer is $x = -1, 0, 2$.

$\Rightarrow g' = f$ changes from increasing to decreasing at $x = -1, 2$ and decreasing to increasing at $x = 0$.

(3) We can infer g graph from g' graph as follows.

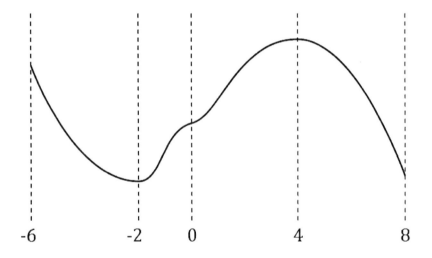

That is, Absolute minimum is either $g(-2)$ or $g(8)$.

$g(8) - g(-2) = \displaystyle\int_{-2}^{8} g'(x)dx \approx \frac{1}{2} \cdot 2 \cdot x + \frac{1}{2}\pi \cdot 2^2 - \frac{1}{2} \cdot 4 \cdot 2 \approx 2 + 2\pi - 4 > 0$

Therefore, from $g(8) - g(-2) > 0$, $g(8) > g(-2)$, therefore the Absolute minimum is $g(-2)$.

From $g(x) = \displaystyle\int_{-1}^{x} f(t)dt$, $g(-1) = 0$.

$g(-1) - g(-2) = \displaystyle\int_{-2}^{-1} g'(x)dx = \frac{1}{2} \cdot 1 \cdot 2 = 1$.

$\Rightarrow$ Therefore, $g(-2) = -1$.

Exercise

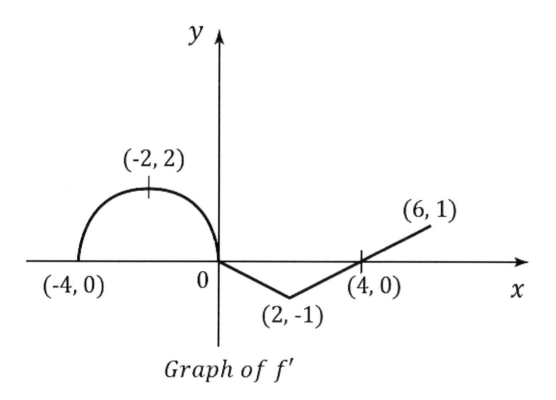

Graph of f'

1. Let f be a function defined on the closed interval $-4 \le x \le 6$ with $f(0)=5$. The graph of f', the derivative of f, consists of two line segments and a semicircle, as shown above.

(1) Find the x-coordinate of each point of inflection of the graph of f on the open interval $-4 < x < 6$. Justify your answer.

(2) Find intervals of f both increasing and concave up.

(3) Find all values of x in the open interval $(-4, 6)$ at which f attains a relative maximum. Give a reason for your answer.

(4) Find $f(-4)$ and $f(4)$.

Solution

(1) $x = -2,\ 2$

f' changes from increasing to decreasing at $x = -2$ and decreasing to increasing at $x = 2$.

(2) $-4 < x < -2$ and $4 < x < 6$

f' is positive and increasing on the intervals $-4 < x < -2$ and $4 < x < 6$.

(3) $x = 0$

f' changes from positive to negative at $x = 0$.

(4)

- $f(4) - f(0) = \displaystyle\int_0^4 f'(x)\,dx$

$\Rightarrow f(4) = f(0) + \displaystyle\int_0^4 f'(x)\,dx = 5 + \dfrac{1}{2}(4)(-1) = 3$

- $f(0) - f(-4) = \displaystyle\int_{-4}^0 f'(x)\,dx$

$\Rightarrow f(-4) = f(0) - \displaystyle\int_{-4}^0 f'(x)\,dx = 5 - \dfrac{1}{2} \cdot \pi \cdot 2^2 = 5 - 2\pi$

Exercise

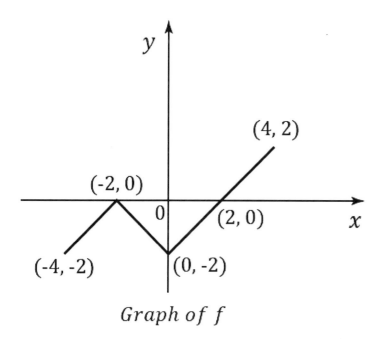

Graph of f

2. The graph of the function f is shown above. Let g be the function given by $g(x) = x + \int_0^x f(t)dt$.

(1) Find all values of x in the open interval $(-4, 4)$ at which the graph of g has a point of inflection. Give a reason for your answer.

(2) Find all values of x in the open interval $(-4, 4)$ at which g attains a local minimum. Give a your reason for your answer.

Solution

$g(x) = x + F(x) - F(0)$

$\Rightarrow g'(x) = 1 + f(x)$

That is, g' graph is the graph that f graph is moved about 1 in the y-axis. Therefore, g' graph is as follows.

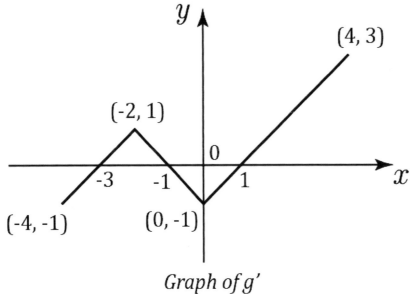

Graph of g'

(1) $x = -2,\ 0$

g' changes from increasing to decreasing at $x = -2$ and decreasing to increasing at $x = 0$.

(2) $x = -3,\ 1$

g' changes from negative to positive at $x = -3$ and 1.

Topic 2

Related Rates

Rate란 짧은 시간 (dt)동안 일어나는 짧은 변화를 말한다.

나음을 보자.

Volume의 변화율이라 하면

$$\frac{dV}{dt} = V'(t)$$ 에서 $dV = V'(t)dt$ 로부터

$$\int dV = \int V'(t)dt \Rightarrow V = \int V'(t)dt$$ 임을 확인할 수 있다.

즉,

$$V(t) = \int V'(t)dt$$

$\uparrow$ Original Value $\uparrow$ Rate

다음의 세 가지 경우를 보도록 하자.

① $\displaystyle\int f'(t)dt = f(t) + C$

② $\displaystyle\int f(t)dt = F(t) + C$

③ $\displaystyle\int F(t)dt = \boldsymbol{F}(t) + C$ (?)

위의 세 가지 중 ③은 애매하다.
그렇다면, 가장 익숙한 모양은 몇 번인가? 아마도 ①번일 것이다.

수학을 잘하는 방법 중 한 가지는 본인에게 익숙한 형태로 "Equation"을 다시 만드는 것이다.

필자는 다음과 같이 제시하고자 한다.

Rate는 무조건 소문자로 $f'(t)$, $g'(t)$ $\cdots$ 등과 같이 표현하자!

예를 들어, rate를 e^{-2t}로 준다면 $f'(t) = e^{-2t}$로, rate를 $P(t) = \sin(t^2)$로 준다면 $P(t) = r'(t) = \sin(t^2)$로 바꾸어 풀자는 것이다.

사소한 이야기 같지만 시험을 볼 때, 상당히 큰 힘을 발휘하게 된다. Rate가 $F(t)$, $P(t)$ 등과 같이 대문자로 표현이 되면 나도 모르게 $\int F(t)dt$, $\int P(t)dt$ 등을 계산하지 못하는 경우가 발생한다.

다음과 같이 알아두자!

Related Rate 문제에서는
1. Rate가 대문자 $P(t)$, $F(t)$로 주어지면 소문자 $f'(t)$, $g'(t)$ 등으로 바꾼다.
2. 완벽한 Equation을 만들자. (생각보다 Equation 만들기는 쉽다.)

보통 AP Calculus AB 또는 BC의 Free Response Question 1번으로 출제되는 경우가 많다.

필자가 제시하는 예제들을 통해 그 풀이를 익혀보도록 하자.

(Example)

An oil tank holds 500 gallons of oil at time $t = 0$. During the time interval $0 \leq t \leq 8$ hours, oil is pumped into the tank at the rate $P(t) = 3t \cdot \sin(t)$ gallons per hour.

During the same time interval, oil is removed from the tank at the rate $R(t) = \sqrt{t}\sin^2(\frac{t}{3})$ gallons per hour.

(1) Is the amount oil on the tank increasing at time $t = 6$? Why or why not?

(2) To the nearest whole number, how many gallons of oil are in the tank at time $t = 8$?

(3) For $0 \leq x \leq 8$, at what time t is the amount of oil in the tank a maximum? Give a reason for your answer.

(Solution)

① Let $P(t) = f'(t)$, $R(t) = g'(t)$, and make equation as follows.

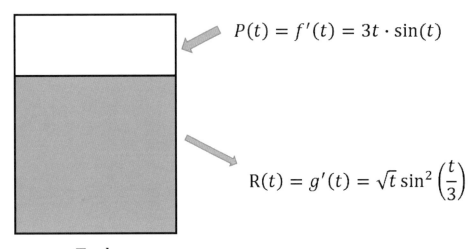

$$P(t) = f'(t) = 3t \cdot \sin(t)$$

$$R(t) = g'(t) = \sqrt{t}\,\sin^2\left(\frac{t}{3}\right)$$

Tank

② Rate of the amount of oil in the tank is as follows.

$$h'(t) = f'(t) - g'(t)$$

(1) No.
$h'(6) \approx -7.055$, and $h'(6) < 0$, therefore decreasing at $t = 6$.

(2) $h(8) - h(0) = \displaystyle\int_0^8 h'(t)dt$

$h(8) = h(0) + \displaystyle\int_0^8 h'(t)dt \approx 496.716 \approx 497$

(3) Draw a graph of $h'(t)$ using a calculator.
From the graph of h', we can get the graph of h as follows.

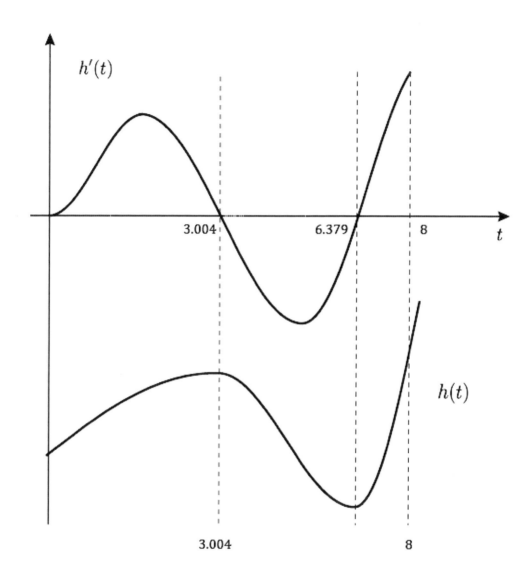

Thus, the candidates for maximum are $h(8)$ and $h(3.004)$.

$$h(8)-h(3.004)=\int_{3.004}^{8} h'(t)dt \approx -11.418 < 0$$

Therefore, $h(8) < h(3.004)$

$\therefore$ h has the absolute maximum value at $t=3.004$

1. For $0 \le t \le 12$, the rate of change of the number of doves at an amusement park at time t months is modeled by $P(t) = \sqrt{t}\sin\left(\frac{t}{3}\right)$ doves per month. There are 300 doves at an amusement park at time $t = 0$.

(1) Is the number of doves increasing or decreasing at time $t = 5$?

(2) How many doves will be at the amusement park at time $t = 12$? Round your answer to the nearest whole number.

(3) To the nearest whole number, what is the minimum number of doves for $0 \le t \le 12$?

Solution

Let $P(t) = f'(t)$. Therefore, $P(t) = f'(t) = \sqrt{t} \sin\left(\dfrac{t}{3}\right)$

(1) $f'(5) = P(5) = \sqrt{5} \sin\left(\dfrac{5}{3}\right) \approx 2.226 > 0$

Therefore, increasing.

(2) Let $f(0) = 300$, then from $f(12) - f(0) = \displaystyle\int_0^{12} f'(t)dt$,

$f(12) = 300 + \displaystyle\int_0^{12} f'(t)dt \approx 309.191$

$\therefore\ 309$

(3) Using calculator, we can draw f' graph as follows.
Find the candidates of absolute minimum, by drawing f graph from f' graph.

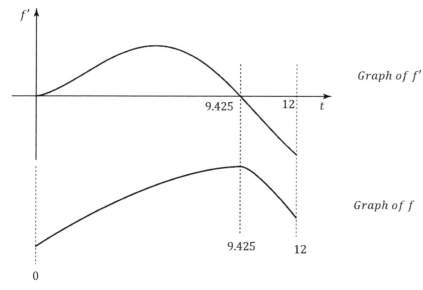

That is, either $f(12)$ or $f(0)$ can be the absolute minimum.

$\Rightarrow f(12) - f(0) = \displaystyle\int_0^{12} f'(t)dt \approx 9.191 > 0$, therefore $f(12) > f(0)$.

Therefore, absolute minimum is $f(0)$.

or

If we find only absolute minimum candidates, we can see from the result of (2).

Therefore, $f(0) = 300$.

Exercise

$$P(t) = 3 \sqrt{t} \sin^2 \left(\frac{1}{2} t \right) \text{ gallons per hour}$$

Oil Tank

$$R(t) = 2 \sin^2 \left(\frac{1}{3} t \right) \text{ gallons per hour}$$

2. In the figure above, an oil tank holds 1500 gallons of oil at time $t = 0$. During the time interval $0 \le t \le 8$ hours, oil is pumped into the tank at the rate $P(t) = 3 \sqrt{t} \sin^2(\frac{1}{2}t)$ gallons per hour. During the same time interval, oil is removed from the tank at the rate $R(t) = 2\sin^2(\frac{1}{3}t)$ gallons per hour.

(1) Is the amount of oil in the tank increasing at time $t = 6$? Why or why not?

(2) To the nearest whole number, how many gallons of oil are in the tank at time $t = 8$?

Solution

Let $P(t)$ as $f'(t)$, $R(t)$ as $g'(t)$, and the rate of amount of oil in the tank as $h'(t)$, then we make the equation $h'(t) = f'(t) - g'(t)$.

(1) $h'(6) = f'(6) - g'(6) \approx -1.507 < 0$

$\therefore$ The amount of oil in the tank is decreasing at time $t = 6$.

(2) $h(8) - h(0) = \displaystyle\int_0^8 h'(t)dt$

$h(8) = h(0) + \displaystyle\int_0^8 h'(t)dt \approx 1500 + 10.172 \approx 1510$

$\therefore$ 1510 gallons

심선생의 주절주절....2

필자는 2008년 Math Level 2책을 처음 출간한 이후로 꾸준히 책을 출간해 오고 있다. 다른
저자분들도 마찬가지이겠지만 책을 출판한다는 것은 살을 깍는 고통을 겪어야 한다. 첫 책을
출판했을 당시 한국에서는 대부분 Math Level 2 시험을 수입서적에 의존을 했었다. 수입 서적이
좋아서가 아니라 공부할 책이 없어서 그랬었는데 필자는 단원 구성부터 모든 것을 필자 주관대로
다시 처음부터 만들었었다. 지금 그때의 책을 보면 "이거 쓰는데 그렇게 힘들었을까?...."라는
생각이 들기도 한다.

이번 AP Calculus AB, BC도 마찬가지로 필자 주관대로 단원을 구성하였고 내용도 필자 주관대로
쓴 부분이 많다. 필자가 3월 AP Calculus Final특강 때 써오던 단원 구성과 강의를 그대로
담아보려고 노력을 하였다. 그러다보니 힘든 부분이 이만저만이 아니였다.

이 책으로 공부하는 학생들의 입장에서는 단원 구성이 생소할 수도 있다. 하지만 믿고
공부해주기를 바란다. 실제 시험은 어느 특정단원을 물어 본다기 보다는 종합적으로 물어보는
문제들이 많다. 그러다보니 필자도 단원 구성을 시험에 맞게 하다보니 실제 학교 교과에서 배우는
단원과 조금은 달라 보일 수 도 있다.

이 책만으로도 충분히 만점이 나오리라 확신이 든다. 하지만 더 공부를 하고 싶은 학생이라면
수입서적을 풀어 볼 것이 아니라 인터넷에 공개되어 있는 최근 주관식시험 몇 해 분과 공개되어
있는 객관식 몇 회분 정도 더 풀어볼 것을 권해 드린다.

Topic 3

Table

우리가 실생활에서 마주치는 실제 Data는 Table로 표현되는 경우가 많다. 그렇다면, 실제 Data를 Table로 표현했을 때, 그 주어진 Data가 "Differentiability"라고 확신할 수 있을까? 또는 식이 주어지지도 않았는데 Integral을 계산할 수 있는가?

그래서, 우리는 Table을 해석할 때, "Differentiable"이라고 Assume을 하여 Mean Value Theorem을 이용하게 된다. Calculus에서 구하는 값들은 Exact Value가 아니고 Approximation이기에 어느 정도 Error가 있더라도 "Differentiable"이라고 Assume하고 계산을 하는 것이다.

다음의 내용을 알고 시험장에 들어가자!

Table이 주어지고 Differentiable이라는 조건이 있는 경우에는 ...

① Mean Value Theorem을 생각하자.

② Integral 계산의 경우 Riemann Sum을 생각하자.

③ Average Value 공식을 이용하자!

$\Rightarrow$ Average Value $= \dfrac{\displaystyle\int_a^b f(t)dt}{b-a}$ $\Rightarrow$ 여기에서 $f(t)$는 Value 자리이다.

예를 들어, Average Volume이면 $f(t)$ 대신 Volume을, Average Velocity이면 $f(t)$ 대신 Velocity가 들어가게 되는 것이다. 다음의 예제를 보자.

(Example)

Height x (km)	0	1	3	5	7	9
Temperature $T(x)$ ($^\circ$C)	40	38	10	8	6	2

The table above gives selected values of the temperature $T(x)$, in degree Celsius ($^\circ$C), x km from the ground. The function T is decreasing and differentiable.

(1) Estimate $T'(6)$. Show the work that leads to your answer.

(2) Write an integral expression in terms of $T(x)$ for average temperature. Estimate the average temperature using a right sum with five subintervals indicated by the data in the table.

(3) Are the data in the table consistent with the assertion that $T''(x) > 0$ for every x in the interval $0 < x < 9$? Explain your answer.

(Solution)

(1) $T'(6) = \dfrac{6-8}{7-5} = -1 \ ^\circ \text{C/km}$

※ The value given in the table is $T = 8$ when $x = 5$, and $T = 6$ when $x = 7$, so for this problem, values like $T'(5.5)$, $T'(6.5)$, $T'(6.7)$ would all have same values.

(2) Average Temperature = $\dfrac{\displaystyle\int_0^9 T(x)\,dx}{9-0}$

It is convenient to draw the figure as follows when calculating the data in the table using Riemann Sum. Height of the square is the Right, so $T = 40$ when $x = 0$ is not calculated.

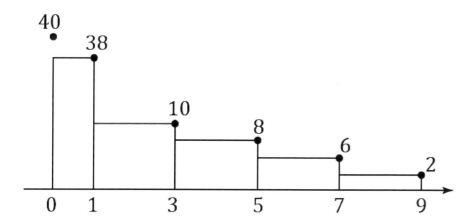

$\dfrac{1}{9}\displaystyle\int_0^9 T(x)\,dx = \dfrac{1}{9}(38 + 20 + 16 + 12 + 4) = \dfrac{1}{9} \times 90 \ ^\circ C = 10 \ ^\circ C$

(3) $T(x)$ is a decreasing function, and $T''(x) > 0$ means concave up. Equation of $T(x)$ is not given, just the data is given by the table. Therefore we can use Mean Value Theorem.

① $T'(x) = \dfrac{38 - 40}{1 - 0} = -2 \ (0 < x < 1)$

② $T'(x) = \dfrac{10 - 38}{3 - 1} = -14 \ (1 < x < 2)$

③ $T'(x) = \dfrac{8 - 10}{5 - 3} = -1 \ (2 < x < 3)$

④ $T'(x) = \dfrac{6 - 8}{7 - 5} = -1 \ (3 < x < 4)$

⑤ $T'(x) = \dfrac{2 - 6}{9 - 7} = -2 \ (4 < x < 5)$

Result of ①~⑤ are as follows.

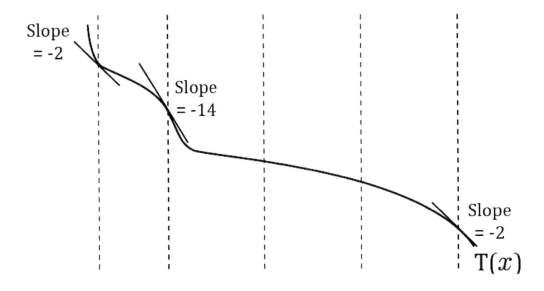

Therefore, $T''(x) > 0$ is not true for every x in $0 < x < 9$.

Exercise

t (minutes)	0	2	4	6	8
$H(t)$ (degree Celsius)	56	50	40	28	14

1. The table above shows the temperature of the potatoes cooled with time when the baked potatoes were taken out of the oven. The temperature of the potato is modeled by a differentiable function H for $0 \le t \le 8$, where time t is measured in minutes and temperature $H(t)$ is measured in degree celsius. Values of $H(t)$ at selected values of time t are shown in the table above.

(1) Use the data in the table to approximate the rate at which the temperature of the potato is changing at time $t = 5$.

(2) Write on integral expression in terms of $H(t)$ for average temperature. Estimate average temperature using a trapezoidal sum with the four subintervals indicated by the data in the table.

Solution

1.

(1) $H'(5) = \dfrac{28-40}{6-4} = -6\,^\circ\mathrm{C/min}$

(2) Average temperature $= \dfrac{\displaystyle\int_0^8 H(t)dt}{8-0} = \dfrac{1}{8}\int_0^8 H(t)dt$

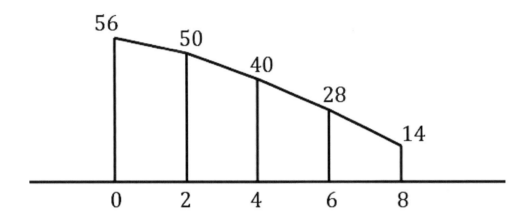

$\Rightarrow \displaystyle\int_0^8 H(t)dt = \frac{1}{2}(56+50)\cdot 2 + \frac{1}{2}(50+40)\cdot 2 + \frac{1}{2}(40+28)\cdot 2 + \frac{1}{2}(28+14)\cdot 2 = 306$

Therefore, $\dfrac{1}{8}\displaystyle\int_0^8 H(t)dt = \dfrac{1}{8}\cdot 306 = 38.25$

Exercise

t (seconds)	0	1	2	3	4
$V(t)$ (kilometer per second)	0	10	30	70	100

2. The table above shows the velocity of the car over time. The velocity of the car is modeled by a differentiable function $V(t)$ for $0 \leq t \leq 4$, where time t is measured in seconds.

(1) Find the average acceleration of the car over the time interval $0 \leq t \leq 4$ seconds. Indicate units of measure.

(2) Write on integral expression in terms of $V(t)$ for average velocity. Estimate average velocity using a right sum with the four subintervals indicated by the data in the table.

(3) Are the data in the table consistent with the assertion that acceleration is positive for every time t?

Solution

2.

(1) Average acceleration $= \dfrac{\displaystyle\int_0^4 a(t)dt}{4-0} = \dfrac{V(4)-V(0)}{4-0} = 25km/\sec^2$

(2) Average velocity $= \dfrac{\displaystyle\int_0^4 V(t)dt}{4-0} = \dfrac{1}{4}\displaystyle\int_0^4 V(t)dt$

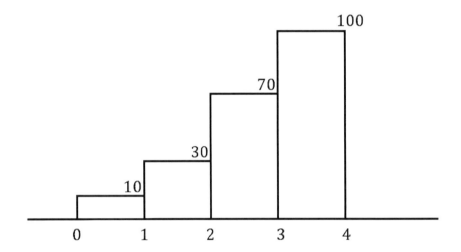

$\displaystyle\int_0^4 V(t)dt = 10+30+70+100 = 210$

Therefore, $\dfrac{1}{4}\displaystyle\int_0^4 V(t)dt = \dfrac{210}{4} = 52.5$

(3) Use a Mean Value Theorem!

$V'(t) = a(t) = 10 \quad (0 < t < 1)$
$V'(t) = a(t) = 20 \quad (1 < t < 2)$
$V'(t) = a(t) = 40 \quad (2 < t < 3)$
$V'(t) = a(t) = 30 \quad (3 < t < 4)$

This can be drawn like as follows.

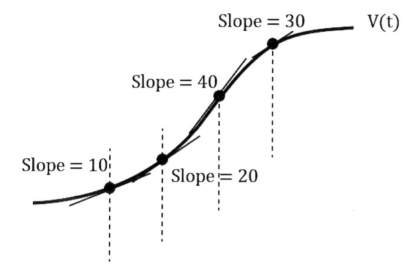

Therefore, the answer is no.

심선생의 주절주절....3

학교에서 Math Course를 계획할 때 남들보다 빠른 것이 좋은걸까?
" 내 친구는 9학년인데 벌써 AP Calculus BC를 듣는다..."
" 저 집 아이는 Precalculus를 시험봐서 건너뛰고 바로 AP Calculus AB반으로 갔데요..."
" 선생님, 제 천재에요. 우리학교에서 10학년인데 벌써 Multivariable을 수강하고 있어요"

이런 말을 들을 때마다 필자는 한숨이 나온다. 모두 그런 것은 아니겠지만 다음 두 학생의 경우를
보도록 하자.

1. Student A
9th Algebra2(Honors) B+
10th Precalculus(Honors) A
11th AP Calculus AB A
12th AP Calculus BC A

2. Student B
9th Algebra2 & Trigonometry A
10th Precalculus(Honors) A
11th AP Calculus BC A
12th Multivariable A

위의 두 학생 모두 필자가 직접 가르쳐왔던 학생들이고 흔히 말하는 미국의 상위 5개 대학에
진학한 학생들이다.

학교 진도가 빠른 것이 보이는가?...아니다. 본인 학년에 맞는 과정을 선택하되 좀 더 도전적인
Honors과정을 선택하여 좋은 성적을 받았다. 흔히 AP Calculus를 일찍 수강하면 수학을 잘하는
학생으로 착각들을 많이 한다. 하지만, 대학에서 원하는 학생은 고교과정을 도전적으로 차분하게
공부하고 올라온 학생을 원하는 것이다. Precalculus를 건너뛸 수 있는 시험을 보고 AP Calculus
AB나 BC반으로 가려는 학생들도 자주 만나 볼 수 있는데 이는 상당히 위험한 생각이다. 왠만한
대학들은 AP Calculus AB 또는 BC에서 5점을 받아 오더라도 대학에서 다시 수업을 듣게 한다.
그러므로, AP Calculus AB 또는 BC 보다는 Precalculus성적을 더 중요시 여기는 대학들도 많다.

 물론 학교코스를 빠르게 잡아서 공부한 학생들 중 명문대에 입학한 학생들도 많을 것이지만
필자가 봐 왔던 학생들은 차분하게 학교의 커리큘럼대로 도전적으로 공부를 해 온 학생들이다.

 물론 교육업쪽에 계시는 원장님들이나 선생님들중에는 다른 의견이 있을 수도 있으니 판단은
본인들의 몫이라는 점을 말씀드리고 싶다.

Topic 4

Area & Volume

1. Area

Area를 구하는 방법에 대해서는 간단하게 소개하고자 한다. 대부분의 학생들이 비교적 쉽게 생각하는 부분이기도 하다.

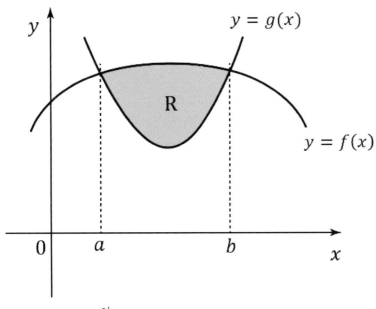

$$\Rightarrow \text{Area } R = \int_a^b (f(x) - g(x))dx$$

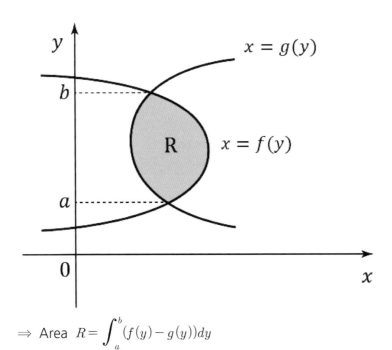

$$\Rightarrow \text{Area } R = \int_a^b (f(y) - g(y))dy$$

2. Volume

AP Calculus에서 공부하는 Volume도 예전에 공부했던 부분과 같다. 단지, 모양이 일정하지 않은 입체도형의 Volume을 정확히 구할 수 없기에 "$\int$"를 이용하여 어느 정도 Error가 있더라도 비슷하게 구하는 것이다.

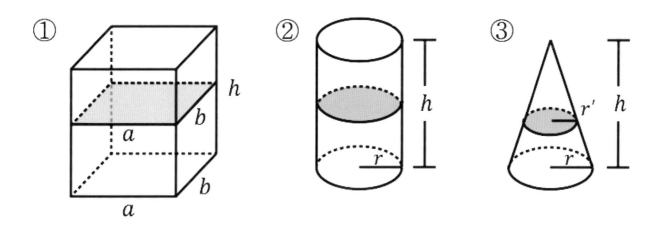

① Volume $= ab \cdot h$

           Height

           Base Area = Cross Section Area

② Volume $= \pi r^2 \cdot h$

           Height

           Base Area = Cross Section Area

③ Volume $= \dfrac{1}{3} \pi r^2 \cdot h$

           Height

           Base Area = Cross Section Area

위의 그림 ①, ②, ③은 AP Calculus를 공부하기 전에 공부했던 내용이다. ①, ②의 경우를 보면 Base Area와 Cross Section Area가 같다는 것을 알 수 있다. ③의 경우 Base Area와 Cross Section Area가 차이는 있지만 앞으로 Integral을 이용하여 공부하는 Volume에서는 같다고 봐도 된다. 그 이유는 다음의 그림을 통해서 확인이 된다.

③

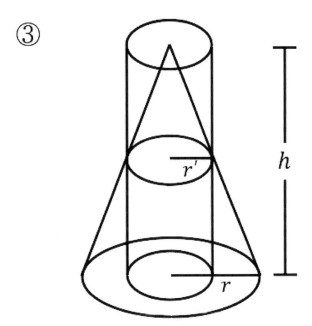

⇒ Cone의 실제 Volume $= \dfrac{1}{3}\pi r^2 h$

⇒ Cylinder의 Volume $= \pi r'^2 h$

Cone과 Cylinder의 Volume은 같지 않지만, 얼추 비슷하게 볼 수 있다. 그러므로, Cone의 Volume도 다음과 같이 계산할 수 있다.

→ **Cone의 Volume** $= \underbrace{\pi r^2}_{\text{Cross Section Area}} \cdot \underbrace{h}_{\text{Height}}$

앞으로 Integral을 이용하여 Volume을 구할 때는 Exact Value를 찾는 것이 아니라 Approximation을 찾는 것이다. 그러므로, 위의 ③번과 같은 계산이 자주 이루어지게 된다.

다음을 보자.

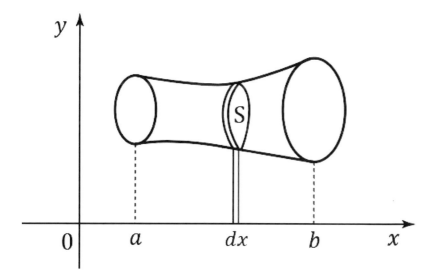

위의 그림에서 주어진 Volume은 다음과 같이 구한다.

$$\int_a^b \quad (of) \quad \left(\!\!\left(\; S \;\right)\!\!\right) \atop dx$$

＝ 많은 합   ＝ Small Volume

그러므로, 다음과 같은 식이 만들어지게 된다.

$$\text{Volume} \;=\; \int_a^b \; S \cdot dx$$

Small Height

Cross Section Volume

우리는 ⟨⟩ 의 Volume을

⟨⟩ 의 Volume으로 대신해서 구하게 되는 것이다.

다음의 그림을 보면서 이해하도록 하자.

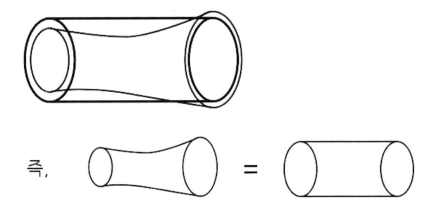

즉,

그러므로, Volume을 구하는 식은 다음과 같이 정해져 있다.

$$\text{Volume} = \int_a^b (Cross\ Section\ Area) \times (Small\ Height)$$
$$= dx \quad or \quad dy$$

여기에서 Cross section의 모양에 따라 다음과 같이 구분된다.

Volume
- ① Cross Section Volume
  - ⇒ Cross Section 모양이 문제마다 주어짐
- ② Revolution
  - Disk, Washer
  - Shell

⇒ Disk, Washer의 경우 Cross Section의 모양이 Circle로 고정되어 있으므로 문제마다 주어지지 않는다. Shell Method에 대해서는 AP Calculus Free Response Question 에서는 굳이 쓸 필요가 없으므로 여기에서는 생략하기로 한다.
본 교재에 자세히 설명해 두었다.

⇒ Disk, Washer의 경우 Cross Section 모양이 왜 Circle일까?
⇒ 이는 당연한 이야기이다.
아무것이나 돌리는 것을 정면에서 보면 모두 Circle로 보이게 되기 때문이다.

다음의 예제를 자세히 분석하면서 Volume을 쉽게 구하는 방법을 연구해 보자.

(Example)

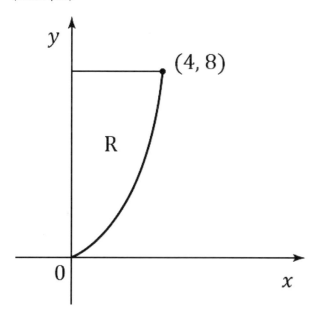

Let R be the region in the first quadrant bounded by the graph of $y = \frac{1}{2}x^2$, $y = 8$, and the $y-$axis, as shown in the figure above.

(1) Find the area of R.

(2) Region R is the base of a solid. The cross section of the solid taken perpendicular to the $x-$axis is a semicircle. Write, but do not evaluate, an integral expression that gives the volume of the solid.

(3) Write, but do not evaluate, an integral expression that gives the volume of the solid generated when R is rotated about the line $y = 9$.

(Solution)

(1) $R = \int_0^4 (8 - \frac{1}{2}x^2)dx$

$\quad = \left[8x - \frac{1}{6}x^3\right]_0^4 = 8 \cdot 4 - \frac{1}{6} \cdot 4^3 = 32 - \frac{32}{3} = \frac{64}{3}$

(2)

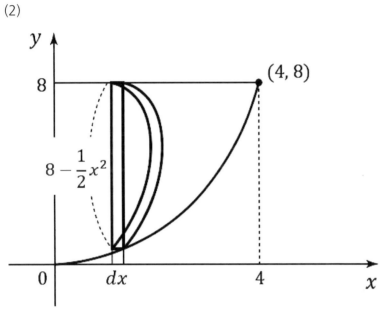

$$\text{Volume} = \int_0^4 (of) \quad y \quad (\text{※ } y = 8 - \frac{1}{2}x^2)$$

$$= \frac{1}{2} \cdot \pi \cdot \left(\frac{1}{2}y\right)^2 dx$$

$$= \underbrace{\frac{\pi}{8}\left(8 - \frac{1}{2}x^2\right)^2}_{\text{Cross Section Area}} \cdot \underbrace{dx}_{\text{Height}}$$

그러므로, $\text{Volume} = \frac{\pi}{8}\int_0^4 (8 - \frac{1}{2}x^2)^2 dx = \frac{256}{15}\pi$

여기에서 원래의 Solid 모양은 다음과 같았을 것이다.

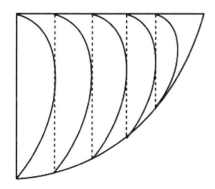

우리가 구한 Solid의 모양은 다음과 같다.

즉, =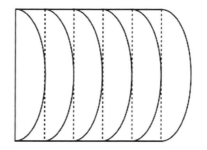

이 문제를 자세히 살펴보면,

① Perpendicular to the $x-$axis이므로 Height가 dx임을 알 수 있고

② Height가 dx이면 Cross Section Side 길이는 y임을 알 수 있으며

③ Cross Section 모양이 문제에서 주어짐으로써 Side 길이만 알아도 우리는 Cross Section의 Area를 알 수 있다.

Cross Section Volume은 다음과 같이 구한다.

① Height는 문제에서 주어진다. (dx or dy)
  - Perpendicular to the $x-$axis $\cdots$ $= dx$
  - Perpendicular to the $y-$axis $\cdots$ $= dy$
  - Parallel   to the $x-$axis $\cdots$ $= dy$
  - Parallel   to the $y-$axis $\cdots$ $= dx$

② Height가 결정되면 Cross Section Side 길이가 결정된다.
  - Height가 dx $\Rightarrow$ Cross Section Side 길이는 y
  - Height가 dy $\Rightarrow$ Cross Section Side 길이는 x

③ Cross Section의 모양이 문제에서 주어지기 때문에 다음과 같이 그림을 그려가면서 구한다.

$$\text{Volume} = \int_a^b (Cross\ Section\ Area) \times (Small\ Height)$$

- $\text{Volume} = \displaystyle\int_a^b ($ **Square**   **Equilateral**   **Semicircle** $\cdots$ $)dx$

  Square: y, $= y^2$
  Equilateral: y, $= \dfrac{\sqrt{3}}{4}y^2$
  Semicircle: y, $= \dfrac{\pi}{8}y^2$

- $\text{Volume} = \displaystyle\int_a^b ($ **Square**   **Equilateral**   **Semicircle** $\cdots$ $)dy$

  Square: x, $= x^2$
  Equilateral: x, $= \dfrac{\sqrt{3}}{4}x^2$
  Semicircle: x, $= \dfrac{\pi}{8}x^2$

(3) 회전축은 항상 x-axis or y-axis가 되게 한다. 회전축을 이동한다 하여도 Volume에는 변화가 없다.

$y = 9$가 회전축이므로 다음의 그림과 같이 전체를 y축으로 -9만큼 이동시킨다.

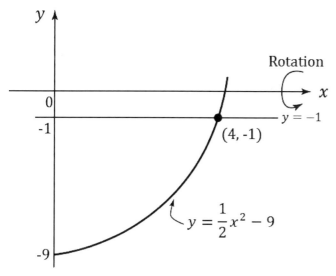

우리가 구하는 실제 Solid의 모양은 다음과 같다.

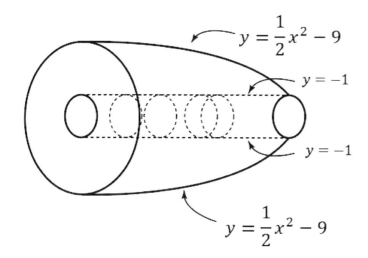

하지만 우리가 실제로 계산하게 되는 Solid의 모양은 다음과 같다.

다음을 보자.

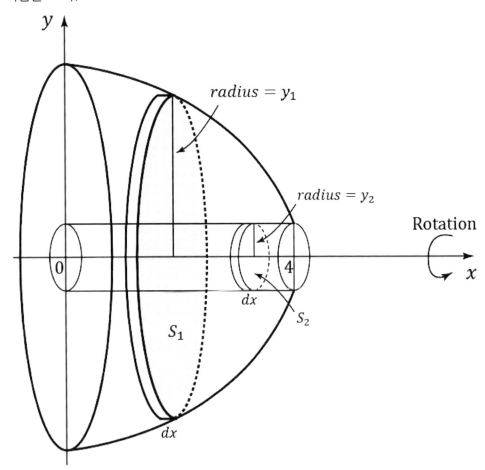

Volume $= \displaystyle\int_0^4 S_1 dx - \int_0^4 S_2 dx$

$\Rightarrow$ S_1과 S_2는 Circle의 Area이므로

Volume $= \displaystyle\int_0^4 \pi y_1{}^2 dx - \int_0^4 \pi y_2{}^2 dx = \pi \int_0^4 (y_1{}^2 - y_2{}^2) dx$

즉, Cylinder - Cylinder가 된다.

그러므로, Volume $= \pi \displaystyle\int_0^4 \left\{ (\frac{1}{2}x^2 - 9)^2 - (-1)^2 \right\} dx$가 된다.

이 문제를 통해서 알 수 있는 사실은

① 회전축은 무조건 $x-$axis or $y-$axis로 해야 한다.

② 회전축이 $x-$axis이면 Height는 dx가 되고 Circle의 radius는 y가 되며

   회전축이 $y-$axis이면 Height는 dy가 되고 Circle의 radius는 x가 된다.

Revolution (Disk, Washer) Volume은 다음과 같이 구한다.

① $x-$axis 회전

$\Rightarrow$ Volume $= \pi \int_a^b (y_1{}^2 - y_2{}^2)dx$

※ y_1은 회전했을 경우 바깥쪽 것이 된다.

만약 y_1과 y_2가 위치가 바뀌게 되면 Volume 값이 Negative가 된다.

이 문제의 경우를 보자.

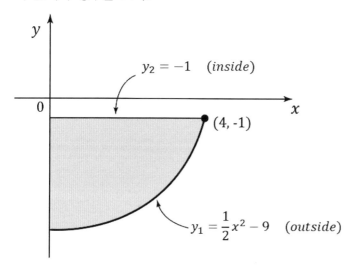

$y_2 = -1 \quad (inside)$

$(4, -1)$

$y_1 = \dfrac{1}{2}x^2 - 9 \quad (outside)$

② $y-$axis 회전

$\Rightarrow$ Volume $= \pi \int_a^b (x_1{}^2 - x_2{}^2)dy$

다음의 예를 보자.

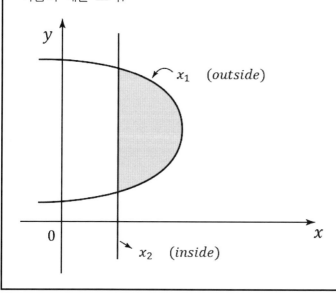

$x_1 \quad (outside)$

$x_2 \quad (inside)$

Exercise

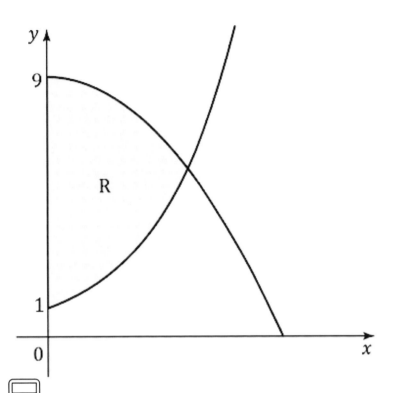

1. Let R be the shaded region bounded by the graphs of $y = -x^2 + 9$ and $y = e^x$ and y-axis, as shown in the figure above.

(1) Find the area of R.

(2) The region R is the base of a solid. For this solid, each cross section perpendicular to the x-axis is a rectangle whose height is 4 times the length of its base in region R. Find the volume of this solid.

(3) Find the volume of the solid generated when R is revolved about the horizontal line $y = 9$.

Solution

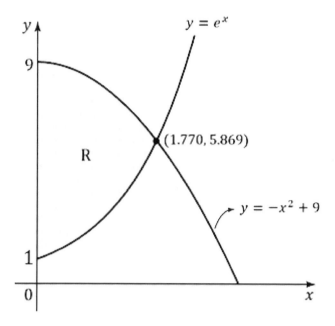

$y = e^x$

(1.770, 5.869)

R

$y = -x^2 + 9$

1.

(1) Area $R = \displaystyle\int_0^{1.770} (-x^2 + 9 - e^x)\,dx \approx 9.211$

(2)

Volume $= \displaystyle\int_0^{1.770} \left(\underset{y}{\boxed{}}\,^{4y} \right) dx$

$= \displaystyle\int_0^{1.770} 4y^2\,dx = 4\int_0^{1.770} (-x^2 + 9 - e^x)^2\,dx \approx 229.338$

(3) If moved to the direction of $y-$axis about -9, the rotational axis would be $x-$axis.

Volume $= \pi \displaystyle\int_0^{1.770} \{(e^x - 9)^2 - (-x^2)^2\}\,dx \approx 216.624$

⚡ Exercise

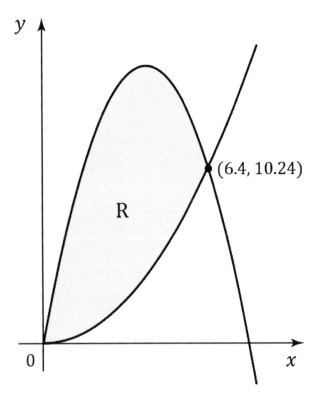

2. Let R be the region in the first quadrant enclosed by the graphs of $f(x) = \dfrac{1}{4}x^2$ and $g(x) = -x(x-8)$, as shown in the figure above.

(1) Write, but do not evaluate, an integral expression for the volume of the solid generated when R is rotated about the x-axis.

(2) The region R is the base of a solid. For this solid, at each x the cross section perpendicular to the x-axis has area $A(x) = \cos(\dfrac{3\pi}{4}x)$. Write, but do not evaluate, an integral expression for the volume of the solid.

Solution

2.

(1) Volume $= \pi \int_0^{6.4} \left\{ (-x(x-8))^2 - (\frac{1}{4}x^2)^2 \right\} dx$

(2) Volume $= \int_0^{6.4} \cos\left(\frac{3\pi}{4}x\right) dx$

심선생의 주절주절....4

학생과 학부모님들 면담을 하다보면 본인들을 자랑하기에 바쁜 학부모님들과 학생들이 있다. 또한 수업을 하다 보면 수업은 듣지 않고 혼자 문제를 풀다가 본인이 모르는 것만 질문하던가 "선생님, 왜 꼭 그렇게 풀어야 하죠? 저는 이렇게 하는데.."..이런식으로 질문을 하는 학생들이 있다. 이런 부분에 대해 필자의 입장을 밝히고자 한다. 물론 다음과 같은 질문은 아주 좋은 예이다. " 선생님, 저는 이렇게도 해 보았는데 어떻게 생각하시는지요? 잘못된 점 있으면 알려 주세요!!"

유명한 축구선수인 손흥민, 메시, 호날두 같은 선수들의 자세를 보면 항상 겸손할 것이다. 본인들은 세계 최정상급 선수들임에도 항상 겸손한 자세를 취하는 이유는 본인들이 배울 것이 더 있다고 생각하기 때문이다. 경기에서 이기고 난 후 항상 그 공을 동료에게 돌린다. 왜일까?.....그저 립 서비스일까? 필자는 아니라고 본다. 더 배우려고 달려들기에 그들은 세계 최 정상급 선수들인 것이다. 이는 필자의 제자들을 통해서도 오랜 기간 입증이 되었다.

수업 중에 수업은 안 듣고 본인것에만 집중하는 학생은 이미 더 이상 올라갈 수 없는 학생이다. 남의 것을 받아들일 준비가 안 되어있고 오직 본인만의 세계에 빠져 있는 것이다. 이미 귀는 닫혀있는 상태로 수업에 참여를 하는데 이와 같은 행동은 학교에서도 이루어 지리라 확신이 든다. 미국교사들에게 이는 상당히 무례한 학생으로 보이기도 한다. 필자가 항상 수업중에 말하는 부분은 " 내가 푸는 방식이 여러분들것과 다른 부분이 있더라도 받아들여라. 여러분들이 공부하던 방식들 위에 더 붙여서 본인들의 실력을 업그레이드 시켜라"

어린아이가 6~7살이 되면 말이 안 통하고 본인의 고집만 강해질 시기이기도 하다. 그 이유는 아이가 아직 배운 것이 없기에 본인이 알고 있는 것이 전부라고 여기기 때문이다. 아이가 크면서 이런 고집이 없어지는 이유는 그 만큼 보고 듣고 배우는게 많아지기 때문이다.

학생들의 경우 평소 독서량이 많았던 학생들이 겸손한 태도를 많이 보이고 수업태도도 좋은 반면 그렇지 않았던 학생들이 수학 좀 배웠다고 본인의 고집대로만 공부를 하려고 한다. 책을 통해 많은 것을 받아 들여봤던 학생들과 그렇지 않은 학생들과이 차이라고 보여진다. 필자에게는 앞에서 말한 어린아이의 경우와 너무 흡사한 부분이 많아 보인다.

부모님들도 마찬가지이다. 우리 아이의 잘난점만 학원에 와서 홍보할게 아니라 아이가 아무리 잘하더라도 아이의 부족한 점 1%를 찾아서 개선시키려 노력을 해야 하는 것이다. 누구나 부족한 점은 있기에...

우연인지는 모르겠으나 필자의 제자들 중 미국 최고의 명문대에 진학한 학생들은 천재들이 아니였다. 모두 겸손한 학생들이었다. 문제풀이 실력이 뛰어난 학생들이 아니였다. 시간관리 철저하고 항상 도전적인 학생들이었으며 주위 친구들 선생들과 잘 어울리는 학생들이었다.

필자가 강조하고 싶은점이 있다.
"누구에게나 꿈은 이루어 진다!! 하지만 그 꿈을 이루기 위해서는 많은 노력도 필요하지만 나에게 부족한 1%를 찾아서 개선켜야 한다!! 1%를 찾아서 개선하려고 노력한다면 본인 스스로 겸손해 질 수 밖에 없다!!"

Topic 5

Motion

Position, Velocity, Acceleration 사이에 다음의 관계가 성립한다. 반드시 암기하여야 한다.

Position, $P(t)$ $\underset{\text{Integral}}{\overset{\text{Differentiation}}{\rightleftarrows}}$ Velocity, $V(t)$ $\underset{\text{Integral}}{\overset{\text{Differentiation}}{\rightleftarrows}}$ Acceleration, $A(t)$

여기에서 t는 time이고 Position, Velocity, Acceleration은 모두 Vector이므로 물체가 오른쪽과 위로 이동할 때가 Positive, 왼쪽과 아래로 이동할 때가 Negative가 된다.

다음과 같이 생각하면 Motion 파트의 문제들도 해결이 쉽게 된다. 이제부터는 Position을 $P(t)$, Velocity를 $V(t)$, Acceleration을 $A(t)$로 쓰기로 하겠다.

$$\underset{= f(t)}{P(t)} \quad \underset{\text{Integral}}{\overset{\text{Differentiation}}{\rightleftarrows}} \quad \underset{= f'(t)}{V(t)} \quad \underset{\text{Integral}}{\overset{\text{Differentiation}}{\rightleftarrows}} \quad \underset{= f''(t)}{A(t)}$$

"The Definition of Definite Integral"로부터

$$\int_a^b f'(t)dt = f(b) - f(a), \quad \int_a^b f''(t)dt = f'(b) - f'(a)$$

를 알고 있을 것이다. 이 식들을 반대로 써보자.

① **Position, Velocity, Acceleration의 관계**

$\rightarrow$

$f(b) - f(a) = \displaystyle\int_a^b f'(t)dt$

$P(b) - P(a) = \displaystyle\int_a^b V(t)dt$

$\rightarrow$

$f'(b) - f'(a) = \displaystyle\int_a^b f''(t)dt$

$V(b) - V(a) = \displaystyle\int_a^b A(t)dt$

② Speed

- Speed $=|$ Velocity $|$

- Speed는 Velocity와 Acceleration의
  작용 방향 (Sign)이 같으면 Increasing 하고
  작용 방향 (Sign)이 다르면 Decreasing 한다.

③ Total Distance

다음을 보자.

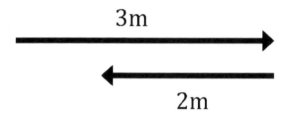

위의 그림에서 Total Distance는 5m가 된다.

$$\Rightarrow \text{Total Distance} = \int_a^b \underbrace{|\text{Velocity}|}_{=\text{Speed}}$$

$$\Rightarrow \text{Total Distance} = \int_a^b (\text{Speed})dt$$

다음의 예제를 자세히 분석하면서 Motion에 대해 연구해 보자.

(Example)

A particle along the x-axis with velocity at time $t \geq 0$ given by $v(t) = -\dfrac{1}{2} + \sin^2(2t)$. At time $t = 0$, the particle is at $x(0) = 3$.

(1) Find the speed at time $t = 2$

(2) Find the total distance traveled by the particle over the time interval $0 \leq t \leq 2$.

(3) Find the x-coordinate of the position of the particle at time $t = 4$.

(Solution)

(1)
$v(2) = 0.073 > 0$
$A(2) \approx 1.979 > 0$

$\therefore$ Speed is increasing since $a(2) > 0$ and $v(2) > 0$.

(2) Total Distance $= \displaystyle\int_0^2 |V(t)| \, dt \approx 0.626$

(3)

$$\underset{x(4)\ -\ x(0)}{\underbrace{f(4) - f(0)}} = \int_0^4 \underset{V(t)}{\underbrace{f'(t)dt}}$$

$$= 3$$

$$\Rightarrow x(4) = 3 + \int_0^4 \left(-\frac{1}{2} + \sin^2(2t)\right)dt \approx 3.036$$

Exercise

1. For $0 \leq t \leq 10$, a particle is moving along the $x-$axis. The velocity of the particle is given by $v(t) = 3\sin(t^2)$. At time $t = 0$, the particle is at $x = 2$.

(1) Find the speed at time $t = 3$?

(2) Find the average velocity of the particle for the time period $0 \leq t \leq 10$.

(3) Find the position of the particle at time $t = 10$.

Solution

1.
(1) Speed $= |v(3)| \approx 1.236$

(2) Average Velocity $= \dfrac{\displaystyle\int_0^{10} v(t)dt}{10-0} = \dfrac{1}{10}\int_0^{10} v(t)dt = \dfrac{1}{10}\int_0^{10} v(t)dt \approx 0.175$

(3) $x(10) - x(0) = \displaystyle\int_0^{10} v(t)dt$

$x(10) = 2 + \displaystyle\int_0^{10} v(t)dt \approx 3.751$

Exercise

2. A particle moves along the x−axis with velocity at time $t \geq 0$ given by $v(t) = e^t + \sin t^2$.

(1) Find the acceleration of the particle at time $t = 2$.

(2) Is the speed of the particle decreasing at time $t = 2$?

(3) Find the total distance traveled by the particle over the time interval $0 \leq t \leq 5$.

Solution

2.
(1) $v'(2) = a(2) \approx 4.774$

(2) $v(2) \approx 6.632 > 0$, $a(2) \approx 4.774 > 0$

$\Rightarrow$ No, since $v(2) > 0$ and $a(2) > 0$.

(3) Total distance $= \displaystyle\int_0^5 |v(t)| dt \approx 147.941$

Topic 6

Differential

Equation

"Differential Equation"의 내용은 다음의 Example을 통해 확인해 보자.

(Example)

Consider the differential equation $\dfrac{dy}{dx} = \dfrac{x}{2y}$

(1) On the axes provided, sketch a slope field for the given differential equation at the twelve points indicated.

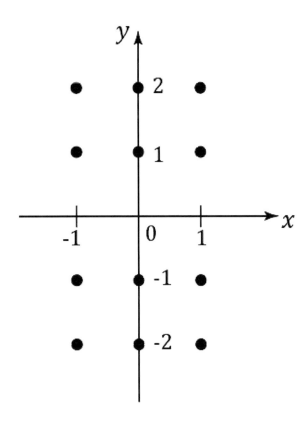

(2) Find the particular solution $y = f(x)$ to the given differential equation with the initial condition $f(2) = -3$.

(Solution)

(1)

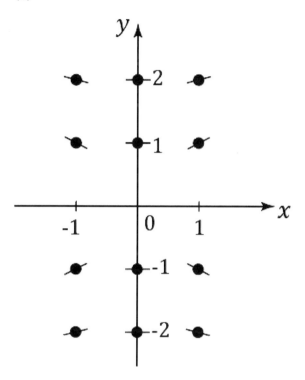

(2)
① 같은 Variable끼리 모은 후 양변에 Integral을 한다.

② $\int 2ydy = \int xdx$

$\Rightarrow y^2 = \dfrac{1}{2}x^2 + C$

(※ $+C$ 는 우변에 한 번만 써준다.)

③ Initial Condition 대입!

$(-3)^2 = \dfrac{1}{2} \cdot 2^2 + C \qquad \Rightarrow C = 7$

그러므로, $y^2 = \dfrac{1}{2}x^2 + 7$에서 $y = \pm \sqrt{\dfrac{1}{2}x^2 + 7}$

④ Initial Condition으로 $+$인지 $-$인지 결정한다.

$-3 = \pm \sqrt{\dfrac{1}{2} \cdot 2^2 + 7}$에서 $-$가 답!

그러므로, 정답은 $y = -\sqrt{\dfrac{1}{2}x^2 + 7}$

Exercise

1. Write an expression for $y = f(x)$ by solving the differential equation $\dfrac{dy}{dx} = x^2 y$ with the initial condition $f(1) = e$.

Solution

$\dfrac{1}{y} dy = x^2 dx$

$\Rightarrow \displaystyle\int \dfrac{1}{y} dy = \int x^2 dx \quad \Rightarrow \quad \ln|y| = \dfrac{1}{3}x^3 + C$

$\Rightarrow 1 = \dfrac{1}{3} + C \qquad \therefore \ C = \dfrac{2}{3}$

$\Rightarrow \ln|y| = \dfrac{1}{3}x^3 + \dfrac{2}{3} \quad \Rightarrow \quad |y| = e^{\frac{1}{3}x^3} \cdot e^{\frac{2}{3}}$

$\Rightarrow y = \pm e^{\frac{2}{3}} \cdot e^{\frac{1}{3}x^3} \quad \Rightarrow \quad y = e^{\frac{2}{3}} \cdot e^{\frac{1}{3}x^3} \quad (\because f(1) = e)$

$\therefore \ y = e^{\frac{2}{3}} \cdot e^{\frac{1}{3}x^3}$

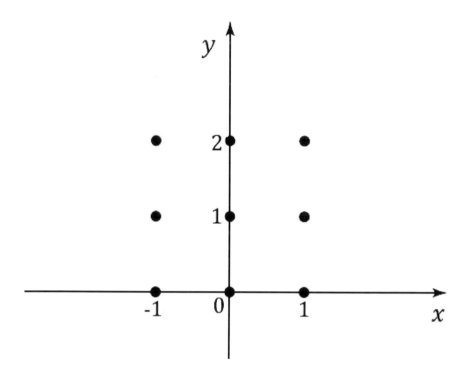

2. Consider the differential equation $\dfrac{dy}{dx} = (x^2 - 2) \cdot (y - 2)$.

(1) On the axes provided, sketch a slope field for the given differential equation at the nine points indicated.

(2) Find the particular solution $y = f(x)$ to the given differential equation with the initial condition $f(2) = 3$.

2.
(1)

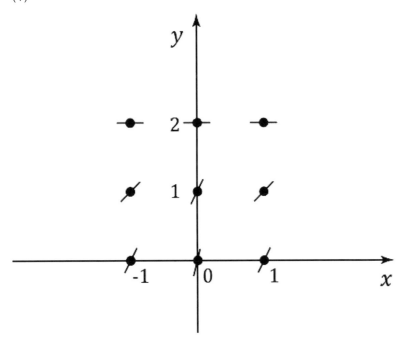

(2) $\dfrac{1}{y-2}dy = (x^2 - 2)dx \quad \Rightarrow \quad \displaystyle\int \dfrac{1}{y-2}dy = \int (x^2 - 2)dx$

$\Rightarrow \ \ln|y-2| = \dfrac{1}{3}x^3 - 2x + C$

$\Rightarrow \ \ln|3-2| = \dfrac{1}{3}\cdot 2^3 - 4 + C \qquad \therefore \ C = \dfrac{4}{3}$

$\Rightarrow \ \ln|y-2| = \dfrac{1}{3}x^3 - 2x + \dfrac{4}{3}$

$\Rightarrow \ |y-2| = e^{\frac{1}{3}x^3 - 2x + \frac{4}{3}}$

$\Rightarrow \ y - 2 = \pm e^{\frac{1}{3}x^3 - 2x + \frac{4}{3}}$

$\Rightarrow \ y - 2 = e^{\frac{1}{3}x^3 - 2x + \frac{4}{3}} \qquad (\because f(2) = 3)$

$\therefore \ y = e^{\frac{1}{3}x^3 - 2x + \frac{4}{3}} + 2$

5월 AP Calculus AB 시험대비
Final 총정리와 Practice Tests

초판인쇄 2019년 4월 5일
초판발행 2019년 4월 5일

지은이 심현성
펴낸이 채종준
펴낸곳 한국학술정보㈜
주소 경기도 파주시 회동길 230(문발동)
전화 031) 908-3181(대표)
팩스 031) 908-3189
홈페이지 http://ebook.kstudy.com
전자우편 출판사업부 publish@kstudy.com
등록 제일산-115호(2000. 6. 19)

ISBN 978-89-268-8782-0 13410